PENGUIN BOOKS

VACATION GUIDE TO THE SOLAR SYSTEM

Olivia Koski was born in the desert and raised in the mountains. She is head of operations for Guerilla Science. Previously she worked as a senior producer at *The Atavist Magazine* and a laser engineer at Lockheed Martin. She lives in Brooklyn with her partner and daughter.

Jana Grcevich lives in Brooklyn and enjoys wandering in Prospect Park. She received her PhD in astronomy from Columbia University and worked at the American Museum of Natural History, where she studied dwarf galaxies and taught astronomy to future high school science teachers.

Guerilla Science creates live, immersive experiences that connect people to science in new and unexpected ways. Since 2008 it has held events for more than 100,000 participants at music festivals, underground bunkers, storefronts, parks, pop-up shops, and more, toward its mission to build a society where science is celebrated as an integral part of culture.

VACATION GUIDE TO THE SOLAR SYSTEM

SCIENCE FOR THE SAVVY SPACE TRAVELER!

Olivia Koski **and** Jana Grcevich

Art by Steve Thomas

PENGUIN BOOKS

for Viola and James, my fellow travel companions
—OK

for my father and mother, who sang me *I See the Moon*
—JG

PENGUIN BOOKS

An imprint of Penguin Random House LLC
375 Hudson Street
New York, New York 10014
penguin.com

ISBN 9780143129776

Printed in the United States of America
10 9 8 7 6 5 4 3 2 1

Set in Archer • Designed by Elke Sigal

CONTENTS

AUTHORS' NOTE

Some may question the wisdom of creating a vacation guide to the planets when human feet haven't touched the ground of another world (the Moon) since 1972. If you're thinking that a space vacation is a distant fantasy, however, remember that one hundred years ago, airplanes were a cutting-edge technology. The fast ones could travel at the "great speed" of 120 miles per hour, bringing a prospective space traveler to Neptune in 2,571 years. In 1989, the *Voyager 2* spacecraft reached Neptune in less than twelve years traveling at 42,000 miles per hour. One hundred years from now, who knows how long a trip to Neptune might take? Your space-vacationing great-grandchildren may discover this book in an old Martian library, and smile at our naive vision of the future.

Assuming we don't destroy ourselves first, humans will go to the places we describe in this book someday, almost without question. With the right resources, and most important the will, we *can* travel to distant worlds. Some of what we discuss, such as human visits to the Moon and Mars, will probably happen in the coming decades. Other things, such as finding ways for our bodies to withstand the extreme radiation near Jupiter, the conditions of the sun-facing side of Mercury, or the long journey to the outer solar system, may take much longer. In some cases we may continue to use probes and robotic explorers to virtually experience distant planets where human survival is impractical.

We are, at heart, space travel agents. Our job is to sell you on the idea of a space vacation. We've created this guide using the best available information on each destination. While we haven't, technically, been to any of them ourselves, rest assured, we have excellent sources who confirm the accuracy of our descriptions of these holiday locations. In this book, we have embellished certain things such as the existence of buildings, cities, and other human-built infrastructure on other planets and moons. Apart from the remains of probes, rovers, and orbiters, six American flags that are likely bleached white on the Moon, and some minor debris, nothing human-made exists on other planetary bodies, and human feet have not touched any alien world except the Moon. Any bit of fiction you do find in this book is informed by scientific and technological expertise, and meant to illustrate the real experience you might have on your vacation.

How can you recognize what is real and what is not? Natural properties of destinations, such as temperature, length of day, climate, etc., are based on the latest scientific research, and we rely on physics to describe how things behave. References to missions, probes, landers, and certain specific rovers are real. The landscapes we portray are real. Names of geographic locations are mostly given according to the International Astronomical Union standards. We chose to use the English translation when possible for readability. The Latin name, commonly used in scientific literature, is often listed in parentheses. Artistic liberties that we took with reality include: references to cities underground and airborne; local rumors;

the possibility of renting rovers, submarines, airships, hover cars, or other vehicles used for navigating alien landscapes; the mere likelihood of surviving exposure to certain types of environments; and the ease with which you can expect to travel to various locations in our solar system.

Creative flourishes aside, we're at the dawn of a space exploration boom. In the short time since Guerilla Science opened its first pop-up Intergalactic Travel Bureau to plan space vacations for the public in 2011, there have been incredible gains in humankind's knowledge of neighboring planets. Scientific missions have brought back images from Pluto, Saturn, and Jupiter, and landed a robot on a distant comet. Thousands of exoplanets, those mysterious worlds orbiting distant stars, have been detected, and more are waiting to be found. The more we discover about other worlds, the more we come to reflect on our own place in the cosmos. Anyone who delves deep into the conditions present on other planets and the extreme measures necessary to keep humans alive on them soon realizes how rare and precious Earth is and how imperative it is that we protect Earth's environment for future generations. The goal of vacationing in space is not at odds with solving our many pressing social, economic, and environmental problems back home. If anything it helps bring into focus the importance of protecting our planet so we have a place to come home to.

As scientists continue making new discoveries about places beyond Earth, and as we struggle to find ways to protect the only planet we know of that is hospitable to human life, entrepreneurs are figuring out ways to make space vacations a reality. Elon Musk's company SpaceX delivers cargo to the International Space Station, and hopes eventually to deliver people to Mars. Other companies, such as Richard Branson's Virgin Galactic and Jeff Bezos's Blue Origin, are racing to be the first to bring private citizens to the edge of space. A company called World View wants to give tourists a view of the curvature of the earth with high-altitude balloon rides, and Bigelow Aerospace, run by budget hotel magnate Robert Bigelow, aspires to rent out inflatable hotel rooms that orbit our planet. These companies are true-life pioneers of the space vacation industry. It's only a matter of time before you, dear traveler, might take a trip.

Your rocket ship awaits.

COUNTDOWN

From the humble ground of your home planet, you can look up at the night sky and see a future of adventure, relaxation, and romance. Every point of light is a possible vacation spot. Where would you like to go?

Your trip will bring you millions or even billions of miles from home. Our Earth-bound understanding of distance is hardly appropriate for making sense of the emptiness that stands between you and vacation bliss. Each journey is different, and we can't predict what yours will hold, but we can promise you will never be the same: Your body, your outlook on life, and your understanding of the universe will be permanently altered.

The places you visit will be both strange and familiar. Earth-centric concepts of space and time give way to the rhythms of a greater physical order. The days may be far longer or shorter. A single year, the time it takes to circle the sun once, may last many human lifetimes. There might not be a ground to stand on. As you scale a towering volcano on another planet, gaze up at the stars from the bottom of a deep crater, or sail through skies filled with strangely colored clouds, the worries you faced in life on Earth will melt away. You'll confront your own insignificance, and smile.

Fear not, we will help you vacation the solar system in style. This is an ordinary guide for an extraordinary holiday. We'll start with the basics: training, packing, and the fundamentals of microgravity health and living. Then we'll get down to the details of your itinerary. We'll cover all of our solar system's planets and other getaway spots, starting with those closest to home and moving to more distant locales. The Moon is followed by Mercury, Venus, and Mars, before we venture to the giant planets of the outer solar system—Jupiter, Saturn, Uranus, and Neptune. Finally, we'll review Pluto. It's no longer considered a planet, but it's just as marvelous a vacation spot as it ever was. In these pages you'll find information on the best time to go, what to expect when you get there, and how to spend your time once you've arrived.

When you're sailing the methane lakes of Titan, rappelling the cliffs of Mars's Mariner Valley, or diving the subglacial seas of distant Europa, you'll finally understand what it means to be an alien. We were built for Earth, after all, and nothing will remind you of your humanness more than an extended vacation to the vast void of outer space.

PREPARING FOR YOUR TRIP

You do not simply decide to take a space vacation and leave the next day. You'll need to train hard, pack light, and get some serious grit. Nothing can *really* prepare you for the feeling of leaving the planet for the first time. But that's exactly why you've decided to go.

Preflight Training

Your body has been shaped by a lifetime on Earth. Getting it ready to weather the new physical and mental sensations of space travel is a full-time job. Embrace the struggle. Assuming you survive the journey, you will cherish the memories you bring back for the rest of your life. A bit of work before departure will help you focus on what's important—relaxation and fun!

Training for your vacation takes months to years, depending on where you plan to go. NASA's qualifications for astronauts are very strict, and unrealistic for the casual vacationer. Let those qualifications be your guide as you consider your trip, and we'll find a way to get you on your holiday regardless of whether you meet them.

- **Vision.** *Visual acuity must be corrected to 20/20 in each eye.* In the past, only those with perfect vision were candidates for space travel. These days, laser surgery makes it possible for a wider group to satisfy this criterion.

- **Blood pressure.** *Not to exceed 140/90, measured in sitting position.* In Earth's gravity, your circulatory system pushes your blood up against the constant pull. In the absence of this downward force, blood can seem like it is rushing to your head. The better your blood pressure is before you go, the better chance you have of avoiding a heart attack brought on by awe-inspiring views.

- **Height.** *Standing height between 4'9" and 6'3".* It's difficult to design one-size-fits-all seating, as tall people who fly in airplanes can attest. NASA's height requirements aren't as strict as they used to be. In the 1960s, astronauts had to be less than six feet tall. Most NBA basketball players would be disqualified from being a NASA astronaut, which is too bad because there's nothing like dunking on the Moon.

- **Military water survival.** Knowing how to survive in water while wearing thirty pounds of combat gear means you'll be ready to face a crash landing.

- **Scuba certification.** Rather than training you to dive coral reefs, this helps prepare you for spacewalks. Learning how to breathe underwater with a compressed air supply will get you used to breathing with air tanks in the vacuum of space.

- **Swim test.** *Swim three lengths of a 25-meter pool without stopping while wearing flight shoes and tennis shoes.* This standard test will determine whether you can meet the physical demands of your vacation.

- **Pressure test.** *Exposure to different pressures in high- and low-altitude chambers.* On your vacation your habitats and space suits will protect you from dangerously high or low pressures. Getting used to big changes in pressure before you go will help you adjust to the disorienting sensations.

- **Gravity training.** *Experience 20 seconds of weightlessness 40 times in one day.* Microgravity environments can be simulated by flying in an airplane through giant parabolic arcs. When the plane flies the downward portion of the arc, you'll spend a bit less than half a minute floating. While on the upswing, you'll experience stronger than normal gravity. If you can survive this dizzying ride on the so-called "vomit comet," you can probably survive everyday weightlessness.

- **Neutral buoyancy training.** Tanks of water are one of the best ways to simulate a microgravity environment on Earth. You'll be fitted with weights to prevent you from floating or sinking.

What to Pack

You've booked your trip and completed your preflight training. Now it's time to pack. Unless you're going for a quick jaunt to the Moon, you need to prepare to leave home for a very long time. Think hard about what you absolutely need to bring, because it won't be cheap to launch. You and your luggage must be shot into space at a minimum of 17,000 miles per hour to get to the first stop on your journey: orbit above Earth's surface.

Back when the U.S. space shuttle was in use, it cost more than $10,000 per pound to get baggage to the International Space Station, in orbit 250 miles above the ground. A $40 airplane baggage fee for a fifty-pound suitcase doesn't seem so bad in comparison. While prices are coming down, even the wealthiest of adventure travelers must respect the laws of physics. Do you need to bring that extra pair of socks? Every ounce you leave home could save you $625.

There are some basics every traveler should consider:

- **Velcro.** After the fiftieth time chasing a loose pen floating through your cabin, you'll appreciate the value of self-adhesive Velcro.
- **Duct tape.** *Apollo* astronauts used this fix-all to repair the fender on their moonbuggy.
- **First-aid kit.** In addition to the basic bandages, medications, and ointments, you'll need supplies for trauma and surgery. Be prepared for anything—you might be the best-qualified surgeon aboard to remove your traveling companion's appendix.
- **Towel.** For spills in microgravity, you'll have to be quick to catch the floating water droplets before they drift away.
- **Soap.** While you won't be bathing very often, if you do bring soap, stick to bars, since liquid soap is messy in microgravity.

- **Face wipes.** While wiping away your facial oils only spurs your face to produce more, many space travelers can't bear to leave their skin unclean.
- **Dry shampoo.** This powdery substance absorbs oil on the scalp without the need to waste water.
- **Clothing.** Stick to garments treated to reduce bacteria, odor, and various skin problems.
- **Camera.** Why go on a space vacation if you can't make all your friends back home jealous with your Moon selfies? Get a ruggedized camera that holds up to harsh conditions.
- **Laptop.** Consider a radiation-hardened model to prevent malfunctions.
- **Toothbrush and toothpaste.** You can bring any old toothbrush, but brushing will be a very different process than you're used to. You'll need to have a drink bag filled with water nearby. Wet your toothbrush; the bristles will suck up the water. Add a little bit of toothpaste, suck some more water from your drink bag, then brush and swallow. Be sure not to let anything float away or escape your mouth during the process.
- **Underwear.** Nothing feels better than slipping on a fresh pair of underwear. Appreciate this simple act as much as possible before you depart, because you'll be changing your underwear a lot less often in space. Japanese astronaut Koichi Wakata wore the same silver-infused, antibacterial underwear for two months on the International Space Station without incident.

- **Pajamas.** A comfortable pair of jammies is a must for temperature regulation and psychological well-being.
- **Activewear.** You'll be working out frequently to avoid osteoporosis.
- **Fancy outfit.** Skirts and dresses are fun but you'll be constantly patting them down like Marilyn Monroe above the subway grating in *The Seven Year Itch*.
- **Jewelry and accessories.** Do pack jewelry that won't float around too much, such as chokers and earring studs, and anklets to show off your shoeless feet. To avoid electrical shorts, don't pack dangly earrings, long necklaces, or anything made from conducting metal.
- **Souvenirs.** Usually you pick up souvenirs while on vacation, but on a space vacation you bring them with you since common items become treasured mementos after they've gone to space. Visitors to the International Space Station are allocated a Personal Preference Kit (PPK), a three-inch-by-three-inch bag for items up to 1.5 pounds. It's just enough to hold jewelry, photos, or patches commemorating your journey. For longer trips, your allowance may be even less.

As you pack, remember that the virtually unlimited supply of breathable air you enjoy on Earth doesn't exist in space. The more gear you bring, the less room there is for the life support systems that provide water, air, and waste processing.

Suit Up

While every traveler's packing list varies, there's one outfit all space tourists will need: a space suit. Space is a hostile place for the human body, and a space suit creates a miniature environment suitable for your physiology. It's a wearable spaceship.

Every suit should come with hydration, temperature regulation, tanks of breathable air, and radiation shielding. A good suit will also be able to withstand minor impacts from micrometeorites, those pesky tiny space pebbles traveling faster than bullets that can cause life-threatening damage.

All this advanced technology is pricey. The cost for a Federal Aviation Administration–approved Extravehicular Mobility Unit starts at around $2 million. NASA spends millions of dollars per year simply to maintain the stock it has on hand. Space suits are modular, so you can share parts with a friend and trade off wearing them to save money. Depending on your itinerary, you may need a separate suit equipped to handle alien atmospheres, temperatures, pressures, and gravities.

A more comfortable—but highly experimental—option is the skintight BioSuit. Instead of filling space between you and your suit with air, it provides comfortable pressure on your skin electromechanically. Advanced fibers woven with electrical circuitry squeeze your body just the right amount as it moves. This gives you maximum mobility in vacuum environments.

When shopping for your space suit, pay special attention to sizing and fit, and test the full suit in a vacuum chamber, which mimics the emptiness of space. While moving around, there should be no pinch points. It should not dig into your armpit the way a crutch might. Places where this happens are called incursion zones. The crotch is the biggest incursion zone, especially for guys. Pay attention to how your suit feels on the back of your knees and crook of your elbows as well, since these can be trouble spots.

The gloves should be good and snug, and your wrists should be able to move freely. Your fingers should reach the tips, and the gloves shouldn't bottom out on the skin between your fingers. Make sure the palm bar, which keeps the glove from inflating into a useless ball, is not pinching you. Performance tests measure how much grip strength, precision, and power you have while wearing the gloves. To get a feel for what these are like, try solving a Rubik's cube with your hands in semi-inflated balloons.

What to Expect on the Spaceflight

You've probably had to endure an airplane flight or two with a crying baby, rude passengers, and cramped seating. A ride on a rocket ship is both more exhilarating and more trying than a jet plane ride. You'll grapple with changes to your body and schedule in new environments with new people, all in the absence of that comforting pull from the planet you've called home for your entire life.

Gravity

From the time you took your first toddling steps, you've had an intuitive understanding of what it means to exist in what we call one g. The constant tug of the earth on you—and you on the earth—is what we think of as "natural" gravity. It's totally reliable. If you get on a scale in New York, you can be certain it will read the same as a scale in Tokyo—give or take, depending on if you snacked on the trip over. That won't be the case in space and on other planets and moons, where you'll be dealing with high gravity, low gravity, microgravity, or artificial gravity.

High g-forces

As you lift off into the sky on your vacation, you'll be pressed back into your seat with an intoxicating force two to three times Earth's gravity. It's like the feeling of being on a spinning Gravitron carnival ride. Most greater-than-one g-forces you experience will be perfectly manageable and kind of fun. However, it's good to know the warning signs when g-forces start to get a little too high.

Your eyes, especially your retinas, are very sensitive to altered blood flow, which happens under high g-forces. Your vision might dim and lose color, called a grayout, like a black-and-white TV show. Tunnel vision comes next, which can progress to an even smaller perspective called gun-barrel vision. In the next stage, blackout, you'll lose your sight entirely, and finally, you'll pass out because of g-force-induced loss of consciousness. However, if the g-force increase is rapid, you might pass out before you notice anything is wrong.

Keep in mind that not all g-forces affect you in the same way. Force exerted from head to toe can be much more damaging than force exerted from the chest to the back. That's why your seatback is parallel to the ground during launch.

You may wish to practice the anti-g straining maneuver in the weeks before your departure. Contract your muscles hard, including your arms, legs, chest, and abs. Take a deep breath, then close off your airway by saying the word "hick." For the next three seconds bear down, then rapidly exhale. Repeat. The maneuver helps to keep blood from pooling elsewhere in the body when it is needed by your brain to keep you conscious in high g-forces.

Microgravity

About ten minutes into launch, the rockets stop firing and you'll transition from feeling as heavy as lead to light as a feather. As soon as it's safe to do so, you'll unbuckle your seat belt and drift out of your seat. Welcome to weightlessness.

Up here in orbit around Earth, you exist in microgravity. This doesn't mean that Earth's gravity fades to nothing as soon as you leave the atmosphere. At 250 miles high, about the distance from New York City to Washington, D.C., the pull of gravity is still 90 percent of what it is on the surface. If you could build a tower that tall and step off, you'd fall to the earth like a rock. But because you and your ship are speeding around the planet at more than 17,000 miles per hour, you're going fast enough to avoid hitting the earth as you fall. Since your spaceship and you are falling together at the same rate, you feel weightless.

Microgravity does strange things to your body. During your first few days in space, you may suffer from "fat face and chicken leg" syndrome. Your cheeks will swell and your legs will become skinnier as your body tries to push fluids up against the force of the gravity it's used to. Good news if you've wanted to grow a few inches—you'll also become taller as the space between your vertebrae expands with fluid.

Low Gravity

Any moon you visit, and several planets, will have lower gravity than Earth. You'll feel light on your feet but a little off-kilter as you learn to walk, run, and generally live life on a world where balls don't fall quite as fast at first and hitting a golf ball four hundred yards is no longer an impressive feat. Ordinary tasks such as juggling and jumping become novel, and on some asteroids and comets you can jump clean off.

Artificial Gravity

Neither the void of space nor other planets or moons ever seem to get gravity right for us humans. High gravity will mess with your senses and possibly cause injury. Lengthy stays on low-gravity moons can weaken bones. In the natural microgravity conditions of open space, two hours of exercise per day is required to avoid muscle atrophy.

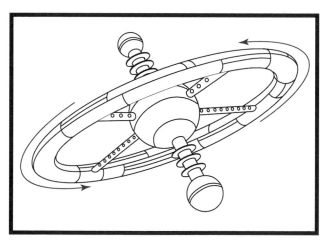

Early NASA concept for a rotating space station

An alternative to all that exercise is artificial gravity. The way it works is simple. You take your spaceship and send it spinning like a giant merry-go-round. The outer edges of the rotating spacecraft become the floor of your artificial gravity home. The strength of it depends on the size of your spaceship and how fast it rotates.

The highest-quality artificial gravity will feel uniform at your head and toes. The best way to get that natural gravity feeling is with enormous ships spinning at imperceptibly slow rates. Smaller (and therefore cheaper) accommodations can make up for their paltry size by rotating faster. This gives many people vertigo. If you don't want your bones to wilt away, you need an occasional dose of artificial gravity—even the budget variety.

SPACE SICKNESS

Virtually everyone who goes to space gets queasy. Even if you've had a stomach of steel sailing stormy seas on Earth, you probably won't escape space sickness. Before launch, make sure you have a little bit of protein in your stomach and are strapped in as snugly as possible. Keep your head still and facing the same direction during your first few days in space. Hold off on the weightless tumbling until your nervous system has had time to adjust. Space sickness pills or patches can also help, as can antihistamines. Don't be ashamed to carry a sickness bag. Nothing is worse than cleaning up floating vomit. Rest assured, you are not alone. Many professional astronauts have hurled in space.

Health

You can't always predict how you'll adjust to new time zones, cuisines, and climates. Seemingly simple tasks such as sleeping, eating, and bathroom time can initially feel overwhelming in environments that disorient the senses. Rest, nutrition, and exercise are essential when a simple spacewalk requires meticulous practice filling air tanks, checking control systems, and inspecting your space suit for leaks. A sickness-induced slip-up can mean the difference between life and death.

Food and Nutrition

Frankly, eating in space is an underwhelming experience. Tastes are muted because of low gravity–induced nasal congestion that interferes with your sense of smell, like eating with a cold. Don't worry, you'll soon get used to eating pouches of rehydratable food. Adding hot sauce to your meals can help add flavor. You might even start to look forward to eating insects. They're an efficient source of protein and can be bred on long-haul space flights. On special occasions you might get a juicy greenhouse-grown tomato or crisp head of lettuce, but most of the time you'll be eating compressed food, which has a high ratio of calories to volume. You'll need to take vitamins to supplement the limited variety in your diet.

Cooking in space is a pain. Open flames and electric stoves are unsafe in low gravity, so you're stuck with microwaves and induction stoves. You'll need to practice rehydrating meals and drinks without making a mess. Crumbs and liquids can float into your eye or get caught in electronics.

On your flight, water for drinking, as well as for washing, brushing your teeth, and even the water that you pass when you go to the bathroom can be processed through a distillation system and used again. Your sweat and breath, which contribute to the ambient humidity aboard your vessel, can be condensed to become a part of the recycled supply of the ship. It's important to check these systems on a regular basis, since no one likes drinking water that tastes like urine.

Sleep

Sleeping in space is an unusual yet restorative experience. Many veteran travelers claim microgravity is the best mattress. You don't need a pillow, and you can rest your head virtually anywhere within your spacecraft. A suitable patch of wall with a sleeping bag and

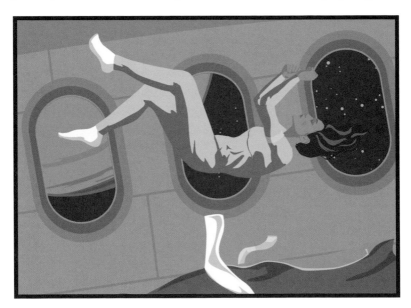

straps to keep you in one place is all you need to settle down for the night. Since there is no up or down you can sleep in any position you'd like. Your arms might naturally fall forward in a zombielike pose, and your head will tip slightly forward. A word of caution to those who fall asleep easily: Be sure that if you're feeling tired you secure yourself to a stable object, or you could float away and bump your head.

As you close your eyes and ease into sleep, you may notice bright flashes. This isn't the paparazzi—it's a cosmic ray zipping through the inside of your eyeball. While it can be disconcerting, it may be comforting to remind yourself that some of these particles have traveled from distant galaxies, only to die inside your head.

A quick note about intimate relations: Your normal routine may become tricky in space. You'll need to figure out the best way to snuggle when the slightest nudge can send you flying. Though many get frustrated and prefer to sleep alone, it's best to think of it as a new and exciting challenge.

Vision

Microgravity has an unfortunate effect on your eyeballs. Your optic nerve swells, pressing against the back of your eye and causing slightly blurred vision. Unlike other changes to your body's shape, your eye can maintain its new form even when you return to Earth. In severe cases your vision may need correction.

Whether you "get something in your eye" or it's the usual existential horror, every now and then everyone needs a good cry. Instead of falling as droplets, the water will build up in your eye. If this happens, don't worry; simply dab away the ball of tears and enjoy the catharsis.

Hygiene

Though you shouldn't sweat much in the filtered air of your ship—maintained at a comfortable 72 degrees Fahrenheit—you'll need to rethink your personal hygiene standards. Clothes don't get dirty as quickly in space as they do on Earth, which is good, because you need to make them last. On any long trip, you will use minimal water or moist towelettes, or simply do without a shower for long periods. If you don't wash, after about five days your clothes become soaked with your secretions and can't absorb any more, and your skin begins to get crusty and rank. Luckily the human sense of smell is very adaptable.

While you will be tempted to change your clothes, doing so only stimulates your skin's oil production. Resist the urge for as long as possible. After about eight days you will adjust and the odor will fade into the background of your perception. Once your clothes are unwearable, simply jettison them, like the astronauts on the International Space Station do.

Toilet

Going to the bathroom in space takes some practice, so brace yourself for a period of toilet training. Low-g toilets employ a form of suction to direct waste safely away. Liquids will go to the water reclamation system and solids will be ejected. If you're growing your own food on a long trip, your solid waste can be diverted for use as fertilizer. You may find that without the help of gravity, things take a bit longer to move than they do on Earth.

Mental Health

Everyone will find their limits tested on an extended voyage in cramped quarters. If you're worried about whether you can hack it, you might want to try an assessment called SUBSCREEN. It's been used by submariners since the 1980s to determine whether military personnel are fit to spend months at the bottom of the ocean. Don't worry, 97 percent of people who take the test pass with no problem. Still, regular mental health checkups are crucial, as the long journey can produce strange effects on your mind.

Radiation

Astronauts on the International Space Station are mostly protected from space radiation by Earth's magnetic field. This radiation can cause damage to your cells by changing your DNA, which can lead to cancer. It might even cause significant brain damage, a condition known as "space brain." Symptoms include anxiety, depression, difficulty making decisions, and memory deficits. If you notice one of your companions starting to act a bit off, it might be time for a checkup. The best way to avoid any possible harm is to make sure you have a well-shielded ship and suit. Or simply stay home.

Risk of Death

There are numerous ways to die in space. Put your affairs in order before leaving Earth.

A partial list of risks:

- **Lack of oxygen.** Your body needs a continuous flow of oxygen, which red blood cells use to produce energy. Though oxygen exists on other planets, it's not in a form you can breathe.
- **Depressurization.** Rapid depressurization, when the pressure drops suddenly, can be fatal.
- **Poisonous gases.** The atmospheres of numerous planets and moons can cause skin irritation or awful burning.
- **Being burned alive.** When fire breaks out on a vessel in space, there is often nowhere to escape to.
- **Falling.** Falling is never a danger in microgravity, where there is no up or down, but you can hurt yourself even when falling in low-gravity environments.
- **Being accidentally left behind somewhere.** Emergency evacuations aren't unheard of, and sometimes the needs of the many outweigh the needs of the few.
- **Running out of food.** Food is tough to grow in space and alien environments, and there is little room for error.

- **Freezing to death/hypothermia.** Though many people associate cold climates with the outer solar system, temperatures can drop rapidly in shady areas of airless environments, even near the sun.
- **Bones being whittled away.** Without constant exercise, your bones will grow steadily weaker.
- **Explosions.** A small explosion can quickly escalate into a ship-destroying disaster.
- **Nuclear incidents.** Nuclear power is what enables you to keep the lights on for those long journeys to the outer planets, and accidents can happen.
- **Asteroid collisions.** Hopefully you'll be aware enough of your surroundings that you'll see asteroids long before they hit. But space is full of surprises.

Destination ☾ Moon

Of the numerous moons of the solar system, there is only one called *the* Moon: Earth's moon. It formed early in our planet's history, when a space rock the size of Mars slammed into Earth. The catastrophic collision thrust molten rock into space, making rings that eventually formed a satellite that's been orbiting Earth ever since.

Despite the familiarity of the Moon in Earth's sky, visitors are shocked to experience its true alien nature. In the words of *Apollo* astronaut Buzz Aldrin, "Nothing prepared me for the starkness of the terrain. It was barren and rolling, and the horizon was much closer than I was used to."

Often the first stop on a longer journey, the Moon serves as an introduction to the strange world of low gravity and the challenges of traveling in the vacuum of space. Highlights include humbling views of the crescent Earth, taking a Moon hopper to visit the historic *Apollo 11* landing site in the Sea of Tranquility, and the sometimes fraught process of learning to walk and play sports in airless terrain where you weigh less than one-fifth what you do back home.

☾ AT A GLANCE

DIAMETER: 25 percent of Earth's

MASS: 1 percent of Earth's

COLOR: Moon gray

SPEED AROUND EARTH: About 2,300 miles per hour

GRAVITATIONAL PULL: A 150-pound person weighs
25 pounds

AIR QUALITY: Traces of helium-4, neon-20, hydrogen,
and argon-40

MADE OUT OF: Rock

RINGS: None

TEMPERATURE (HIGH, LOW, AVERAGE): 240, -290,
-4 degrees Fahrenheit

DAY LENGTH: 708 hours and 54 minutes

YEAR LENGTH: 1 Earth year

TRAVEL TIME AROUND EARTH: About 27 Earth days

AVERAGE DISTANCE FROM THE SUN: 93 million miles

DISTANCE FROM EARTH: 222,000 to 253,000 miles

TRAVEL TIME: 3 days

TEXT MESSAGE TO EARTH: 1.3 seconds

SEASONS: Very mild

WEATHER: None

SUNSHINE: About the same as on Earth, but harsher light

UNIQUE FEATURES: Tycho Crater in the southern hemisphere

GOOD FOR: A quick getaway

Weather and Climate

Since there is no atmosphere on the Moon, there is no weather, and many visitors find the quiet surroundings rather relaxing. You won't need to worry about seasons, either. The Moon's axis barely has any tilt—1.5 degrees—so the amount of sunshine you get stays the same throughout the year, no matter where you are. Though you won't encounter any unexpected storms, the temperatures fluctuate wildly, from 240 degrees Fahrenheit during the day to -290 at night, making it rather bothersome to pack for outings. It's like preparing for an excursion to Earth's hottest desert with an overnight in the Antarctic. Fortunately this dramatic temperature swing only occurs every fourteen Earth days, so you have time to adjust to the new conditions. Though space is by nature a frigid place, and anywhere you go you're likely to encounter record-breaking cold snaps, on the Moon it can get as cold as it does anywhere else in the solar system: -400 degrees Fahrenheit. It's Pluto cold. Those who love this numbing weather can find it at the Moon's south pole, in craters that are so deep that the sun's rays can't reach them.

If you prefer moderate temperatures, the best time to venture outside is during lunar dawn. Just be careful—at dawn, the peace on the Moon is sometimes disrupted by moonquakes, which are triggered when the cold crust is warmed by the sun for the first time in two Earth weeks. These and other quakes that start deep below the surface—or those caused by meteorite hits—are mostly mild and harmless. It's the shallow ones ten or twenty miles underground that can rattle heavy furniture and shake buildings. Because the Moon is so dry and frigid, it rings like a bell, and quakes can last up to ten minutes. If you find yourself caught in one, don't panic. Stay calm and try to enjoy the ride.

To get the full Moon experience, be sure to stay a full lunar day. It's longer than it sounds—a day on the Moon lasts almost thirty Earth days. That will give you plenty of time to explore both the near and far sides.

When to Go

If you haven't yet been to the Moon, you should go immediately. Close this book, call your local intergalactic travel agent, and reserve a trip. What are you waiting for? There's no time like the present, since the Moon is drifting away from Earth at a rate of one and a half inches a year. Wait eight years and you'll have to travel an extra foot.

Going to the Moon is like traveling around the world ten times, since that's how far you will have traveled once you arrive. Rockets travel far faster than planes, and you'll be there in just a few days, marveling at its gray surface from seventy miles away. If there's a place you want to see while the sun's out, plan ahead since the lunar night lasts so long. If you stay a whole Earth month, you're guaranteed to get a sunlit view of your favorite spots. Check the illumination conditions against your itinerary.

Getting There

Your journey to the Moon begins at a spaceport. Spaceports are just like airports. In most cases, you'll depart from a launchpad, rather than a runway. They tend to be in the desert or near bodies of water, just in case a rocket explodes or crashes at takeoff. You can take a preholiday outing near your launch site before you depart, since spaceports are built in areas known for clear skies and calm weather. These may be your last memories of Earth.

Breaking the bond of Earth's gravity will require you to attain the great speed of 25,020 miles per hour, known as Earth's escape velocity. To achieve this speed, you will be strapped to a (hopefully) controlled explosion. A single launch from Earth takes almost half the energy that a flight all the way to the edge of the solar system requires. This is the origin of the old spacefarer's saying, "Half the journey is getting to orbit." You'll save money by choosing flights that depart near the equator. Taking off from there on an eastward launch gives your rocket a kick from the spin of the earth.

Jet engines are great for taking you from one side of the earth to the other, but they require something in short supply in space: air, specifically the oxygen found in air. Chemical rocket fuel that powers many rocket ships provides its own oxygen to create thrust. For a short trip to the Moon, you don't have to worry about fueling up along the way. If you're headed to more distant settings, the basic ingredients for chemical rocket fuel are available on many terrestrial worlds, which means you don't have to pack all your fuel with you.

The computers that brought the first visitors to the Moon were much less powerful than your smartphone. The 240,000-mile journey takes three days, but if you're just passing by without stopping, you can get there in as little as nine hours. It's hard to get lost on your way to the Moon, since you can see it from Earth and it's a simple matter of keeping your ship pointed in the right direction.

En route, watch out for the Van Allen radiation belts, which are zones of trapped particles. They can wreak havoc on electronics, though they appear to cause little harm to humans. There are two main sections: One is 400 to 6,000 miles above Earth's surface, while the other is between 8,400 and 36,000 miles from it. *Apollo* mission scientists were initially concerned they might cause health problems for astronauts, but onboard radiation detectors showed that radiation remained at safe levels during the crossing. You might want to see which of your travel companions can hold their breath the longest as you sail through.

Once you've reached orbit around the Moon, take part in the traditional celebration of cracking a bottle of champagne. Just be

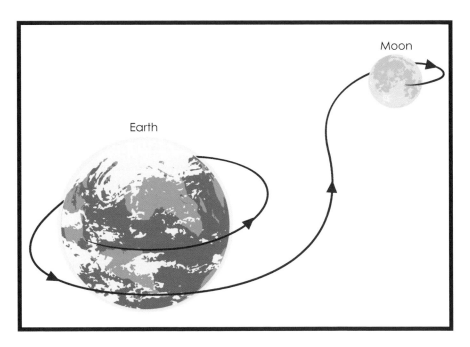

The Moon is just a short few days' trip from Earth.

careful, as opening a bottle can be a dangerous endeavor in the low-pressure environment of your spaceship. Watch out for the cork as it shoots out of the bottle at speed, as well as the inevitable floating blobs of bubbly.

When You Arrive

Many find that during their first close encounter with the Moon they experience a delightful combination of recognition and otherworldliness. The familiar features of the man in the moon resolve slowly into vast dark plains, expansive craters, and mountain ranges. Upon arrival, some visitors settle into a low orbit around the Moon's equator. Don't rush to reach the surface; many sights are best seen from far above, and you'll have plenty of time to explore them up close.

Others will opt for the most economical option for reaching the surface: being lowered by a space elevator. You'll start far from the Moon at a place called L1, or lunar Lagrange point 1. At L1, which is between the orbits of Earth and the Moon and closer to the Moon, the combined gravity of Earth and the Moon creates an orbit that stays in a fixed position with respect to the Moon. The lunar space elevator connects a space station at L1 with the surface via a superlong, strong cable. You'll enter a capsule with your belongings and will be slowly lowered to the surface. While more efficient and therefore less expensive than rocketing down, space elevators can be damaged by micrometeorite impacts.

If you're arriving on the lunar near side, don't forget to turn back and view your home planet. Gazing at the fragile blue marble suspended in the darkness, first-timers often experience feelings

A snapshot of everywhere you had ever been before your trip.

NASA/GSFC/ARIZONA STATE UNIVERSITY

of intense awe and wonder. Keep your wits about you enough to take a photo, perhaps looking like you're holding the entire world in your hands.

Be sure to request an Earth-view room when you check into your hotel. This allows you to view the beauty of Earth when it is fully lit, as a shimmering crescent, and as a dark circle against the twinkling stars. Because the same side of the Moon always faces Earth, the planet will never move out of your window. Reports of lunar werewolves, which are said to come out only during the full Earth, are greatly exaggerated, as are rumors of roving serpents with many heads called the "Pahasydämiset." (That story is thought to have originated from an obscure Finnish folktale, and is a fabrication as far as we know.) During the new Earth, look for the lights of civilization and the outline of continents in the shadows on Earth.

As you get settled, you'll want to familiarize yourself with tourist etiquette and protocol. Always keep building airlocks unlocked; it's considered a life-or-death matter, a necessary convention when a small hole in your suit can endanger your life if you're stuck outside. As water is scarce, wasting even small quantities of it is seen as an unforgivable offense. And watch out for anyone trying to sell you land on the Moon. Despite what they claim, land cannot be bought or sold here, and anyone trying to sell it to you is attempting to trap you in a real estate scam.

Getting Around

Freedom from strong gravity brings freedom from traditional methods of overground travel. Anyone can rent a rover, but those who are really in the know travel by hopper. Hopping the Moon is not only an unforgettably fun experience but also a very efficient method of long-distance moon travel. Hoppers look like the original *Apollo* lunar lander, with four legs and an area for seats, cargo, and fuel. They are equipped with hinged legs and springs, and their power sources build up energy, which is released suddenly in one dramatic bound. With no air resistance, you'll sail over the surface in a large arc, covering distances as short as four hundred feet and as long as three hundred miles in one jump. The speed at which you travel over the surface will be determined by how much power you hop off with, as well as the local surface gravity. For long distance trips you'll need to chart out pit stops to fill up your fuel cells with hydrogen and oxygen.

If you prefer to travel by rover, expect to inch across the surface at a leisurely golf cart speed. Lunar rover speedometers max out at 12 miles per hour. If you traveled at 10 miles per hour without stopping, you could circumnavigate the Moon in about twenty-eight days.

Airplane travel isn't possible because of the lack of atmosphere. Any flying you do requires rocket power. When you're ready to depart, getting back into orbit isn't exactly cheap. You'll need to reach a speed of 5,300 miles per hour to escape the pull of the Moon's gravity.

What to See

The Near Side and Far Side

Despite Pink Floyd's claim, there is no dark side of the Moon. Like on Earth, everywhere on the Moon has a regular cycle of daylight and darkness, except for the poles. During the day, the sun is bright and the sky is black, but you won't see many stars. At lunar night you won't see the sun, but the rest of the stars in the sky will be brilliant.

Though there is no permanently dark side of the Moon, there is a far side that never sees Earth because of tidal locking. Earth's gravity pulls unevenly on the Moon, from its front to its back. This stretches out the Moon, making it bulge on both sides. Over time, Earth's pull on the bulges has slowed the Moon's rotation so only one side is visible from Earth. Like a friend trying to hide

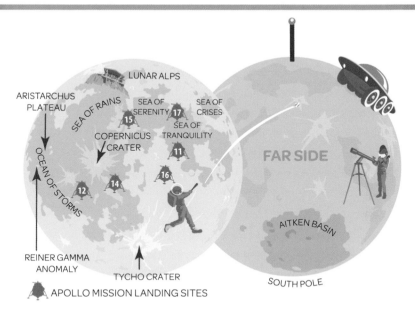

Both the near and far sides of the Moon have plenty of amusements to offer first-time visitors and veterans alike.

something behind him or her, the Moon scooches across Earth's sky, never showing its backside. Before the Moon became a hot vacation spot, there were wild rumors about what might be hidden on the far side, and it maintains a certain air of mystery to this day.

You'll notice the near side has many more flat areas, or *maria*, Latin for "seas," than the far side. Just after the Moon formed, Earth kept the near side warm and molten beneath its surface. When asteroids bombarded the near side, lava broke through and cooled into plains. The patterns of *maria* form the face of the man in the Moon. You can finally enjoy his features up close.

Lacking the *maria* of the near side, the far side has craters everywhere. According to *Apollo* astronaut William Anders, the far side of the Moon "looks like a sand pile my kids have been playing in." Its rumpled appearance is not because of its exposure to more space debris than the near side, but because its molten surface hardened more quickly than the near side, leaving older craters preserved for your enjoyment. You won't get views of Earth from the far side, and that's exactly why some people like to go.

Historical Points of Interest

History buffs will enjoy the six *Apollo* landing sites, particularly the place where *Apollo 11* landed in the Sea of Tranquility. Be sure to visit the perfectly preserved first footprints on the Moon, left in 1969 by astronaut Neil Armstrong. Amazingly, the Moon's first visitors traveled 238,900 miles to walk in circles no bigger than a baseball field. They stayed for just twenty-two hours, going outside for two and a half hours to kick dirt, practice walking, and collect rocks to bring back home. The flag they planted, which fell down as they left, has been bleached white from sunlight and radiation. Subsequent *Apollo* mission visitors had more opportunity for tourist activities, such as golfing.

The Chinese moon program, named Chang'e, after the Moon goddess, provides another historic landmark. A rover from the program, named Yutu (Jade Rabbit), after Chang'e's pet, explored the surface for thirty-one months before it live-blogged its own death. Its final message was "Goodnight, Earth. Goodnight, Humanity." You can visit its final resting place in the Sea of Rains (Mare Imbrium).

The Moon Museum

The Moon Museum is a tiny ceramic wafer, three-quarters of an inch by half an inch in size, that contains minuscule black-and-white art works by six twentieth-century pop artists: John Chamberlain, Forrest Myers, David Novros, Claes Oldenburg, Robert Rauschenberg, and Andy Warhol. This subversive "museum" was smuggled aboard the spacecraft for the *Apollo 12* mission and ferried to the Moon without official approval from NASA. As far as anyone knows, it remains among the *Apollo 12* ruins to this day.

Ocean of Storms (Oceanus Procellarum)

The Ocean of Storms is the only *mare* vast enough to be named "ocean" rather than "sea." After visiting the *Apollo 12* and *14* landing sites, head over to the mysterious Reiner Gamma anomaly. This forty-four-mile-wide geologic feature is a bright white splotch resembling something between a tadpole and cream being poured into coffee. The "Area 51" of the Moon, it's home to a strange disturbance in the Moon's magnetic field. The magnetic field strength is thought to contribute to the formation and preservation of the feature by affecting the amount of radiation reaching the surface. Reiner Gamma is also known as one of the largest sources of "transient lunar phenomena." These are strange flashes of light, changes in color, and other short-lived bursts of activity witnessed by Earth observers for at least the past one thousand years. Though it's impossible to confirm which events really took place, scientists think some flashes are related to gas escaping from lava tubes and other geologic activity in the area.

Enjoy the mysterious swirling patterns of Reiner Gamma, an area of bright and dark ground thought to be related to magnetic effects.

NASA/GSFC/ARIZONA STATE UNIVERSITY

Before you leave the Ocean of Storms, stop by Copernicus crater, located to the east. It has dramatic central mountains, and its walls are complex and terraced. Although it is thirty-seven miles wide, it is not very deep. If you scaled Copernicus to the size of a nine-inch pie pan, it would only be one-third of an inch deep.

Aristarchus Plateau

To the north of the Ocean of Storms, Aristarchus plateau, also known as Wood's Spot, was named after the well-known Greek astronomer Aristarchus of Samos, who was the first to widely propose a model of the heavens with the sun at its center. The Cobra Head, in the central region of Aristarchus, is a boulder-rich region with splashes of light and dark rock flowing down its slopes. Extending to the side of the plateau and beginning with the Cobra Head is Schröter's Valley, a snakelike eighty-seven-mile-long valley formed by lava.

No trip to the region would be complete without a visit to the famous Aristarchus crater, whose brightness pierces a particularly dark area of the plateau. The crater has rich concentrations of the mineral ilmenite, which can be mined for oxygen used in rocket propellant.

Eastern Sea (Mare Orientale)

West of Aristarchus, spanning the border between the near and far sides of the Moon, is the Eastern Sea. It resembles a bull's-eye of bright and dark rings. The structure was created when a large impact crater flooded with lava, then cooled and hardened to a smooth surface. These ripples are concentric mountain ranges; the innermost is called the Inner Rook Mountains, followed by the next ring, the Outer Rook Mountains, and finally the Cordillera Mountains.

Though the terrain is rugged, many come to the Eastern Sea for a spiritual pilgrimage to the center of the bull's-eye, starting outside the Cordillera Mountains and moving inward. Because of the lack of atmosphere, there is no sound, which enhances the silent experience for meditators.

Tycho Crater

Tycho is the Moon's most recognizable feature and crowded tourist spot. This circular blemish in the southern hemisphere is highlighted all around by brilliant rays of disrupted lunar rock, and is easily seen from Earth. The crater is so popular that it's referred to as "the metropolitan crater of the Moon," a phrase coined in 1895 by author Thomas Gwyn Elger. It was formed in an asteroid impact about 109 million years ago. Had you been around during the time of the dinosaurs, you could have seen a bright flash in the sky from Earth as it was hit.

Settle in on its northern rim, where you can relax and enjoy views of the fifty-three-mile-wide pit. The floor is the perfect place for a fifty-mile ultramarathon. The rim of the crater is dark, an artifact of lunar rock that melted during a meteor impact and resolidified into a thin glassy layer. You can collect pieces of moon glass from the ground to bring home for souvenirs. Many visitors choose to hike to its base along its zigzagging slopes, three vertical miles from the top. A zip line on the eastern rim could get you down in about an hour.

Ziplining on the Moon is a leisurely affair because of low gravity; your ride will have a gentle start. You don't need to worry about too much wind in your face or running into bugs, since you won't find either on the Moon. Once you've made it to the bottom, it's time to start climbing again. The peak

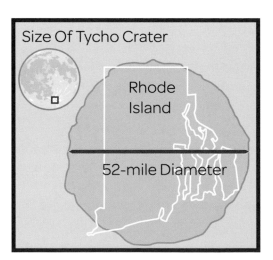

The footprint of the prominent Tycho Crater compared to the size of the state of Rhode Island.

in the middle of Tycho's base provides broad views of the crater and surrounding terrain along its rim.

South Pole

After visiting crowded Tycho, you may choose to journey south to Aitken basin. Covering a good portion of the far side's southern region, it's one of the largest craters in the solar system, with an abrupt eight-mile drop and a base stretching for sixteen hundred miles.

Within the basin are the Leibnitz Mountains. The highest peak in the range, at about twenty-seven thousand feet, is a couple thousand feet shy of the height of Mount Everest. From its steep cliffs, you can see the entire southern polar region. You must train before attempting to summit the giant slope, though in the low gravity it's easier to get to its peak than that of its sister mountain on Earth. Altitude sickness, which plagues hikers who summit high mountains on Earth, is not a risk on the Moon where everyone is required to carry an ample supply of breathable air.

After a grueling hike to the top of the Leibnitz Mountains, you may want to take a respite in one of the pole's permanently shaded regions. These areas are located at the bottom of certain craters, which escape exposure to sunlight and are highly prized for their extremely cold, dark, and stable environments. They are known as "craters of eternal darkness." Shaded regions can be used to store propellant and other supplies. They make ideal crypts for the bodies of people who have arranged to be cryogenically preserved after death, since they provide eternal natural refrigeration.

While you're in the south, don't miss Shackleton crater. While its bottom is a permanently shaded region, its rim is home to peaks of eternal light, where the sun always shines. If you want to explore the smaller craters and peaks within it, you'll have to take a hopper because rovers can't manage the steep drop to the bottom.

The Lunar Alps and Sea of Rains

The lunar Alps, located on the northeast edge of the Sea of Rains, have their own unique brand of sparse beauty. Begin your journey at the ancient crater Plato to the north of the Alps, which has a collection of tiny, more recent craters in its basin. From there, drive your moon buggy southward to the majestic Alps themselves, named Montes Alpes in Latin. Next, visit Vallis Alpes (Alpine Valley), which slashes through the middle of the mountain range, connecting the Sea of Rains to the Sea of Cold.

One of the highest peak in the lunar Alps, Mont Blanc, is named with a French flourish. While Earth's Mont Blanc lies on the border of France and Italy, the lunar Mont Blanc lies just south of the western end of Alpine Valley. Its peak rises about 12,500 feet above its surroundings. Although Mont Blanc's slopes are not steep, it will still take a long Earth day's climb to get to the top. Since the sun is up for fourteen Earth days at a time, you should be able to get there well before lunar sunset. From Mont Blanc we recommend a slight jaunt to the west to visit Piton Mountain (Mons Piton), a small but conspicuous peak located in the flat plains of the Sea of Rains. Once you've practiced climbing there, head south to the Apennine Mountains (Montes Apenninus) and the highest mountain on the Moon, Mount Huygens (Mons Huygens). The mountain rises 3.3

miles from its base and will give you a chance to hone your lunar mountaineering skills.

Activities

Experience the "Joys" of Moondust

Moondust may have a romantic name, but it's like a stray puppy: It will follow you everywhere and never leave you alone. Called regolith, this fine dust covers most of the Moon's surface, sometimes to a depth of forty feet. During your time visiting the Moon, you'll get to know moondust intimately. It gets in everything. As you sweep the floor yet again in a futile effort to keep things clean, whistle. Re-

MOONWALKING ON EARTH

| Turn on some Michael Jackson and start with your feet together. | Lift up the heel of your left foot and put your weight on your toes. | Slide your right foot backwards, keeping it flat with the floor. | Raise your right heel. Move your left foot backwards. | Switch feet and repeat. |

WALKING ON THE MOON

| Push off one leg very gently. Take a long step with your leading leg. | Let your foot bounce off the surface without bending it too much. Alternating legs is optional. | For faster but less controlled motion, consider hopping higher. | When you need to stop or change direction, lean back and dig in your heels to generate enough friction to stop your forward motion. |

45

member: You get to sweep a floor on *the Moon*. When you kick it up, it will fall slowly, like snowflakes made of dust bunnies.

Regolith is very useful. You can make lunar concrete by heating it with energy from the sun or microwaves in a process called sintering. It's used to make roads and buildings. You'll soon notice the same sophisticated shade of moon gray everywhere, similar to the color of asphalt. Like regular old Earth concrete, it provides excellent protection from solar radiation—necessary for survival on the Moon.

Walk on the Moon

Walking on the Moon is like walking on a trampoline underwater. It takes at least ten minutes to get the hang of the basics, but it can take years to override your earthly intuition and master lunar locomotion. When walking on Earth, we fall forward while our weight is on one leg. If you tried to walk this way on the Moon, you'd find it very slow, inefficient, and unstable. Because your muscle power does more on the Moon, you don't need to bend your legs as much to walk. You can easily overdo the amount of force you push off with, and low gravity may cause you to go flying upward, unable to stop or control your motion.

A favored technique is to alternate leading legs and to use longer strides, much like a slow cross-country skiing motion; Neil Armstrong called this method the "lope." An alternate approach is to take shorter strides, keeping one foot on the ground at all times in a sort of shuffle while leading with the same leg.

Spaceball, the Moon's planetary pastime

Play Spaceball

Exercise in the low-gravity environment of the Moon is a blast, once you get used to it, and it has the added benefit of keeping your bones and muscles from becoming delicate. Be forewarned: Though you'll feel powerful with your seemingly superhuman strength, you can seriously injure yourself without proper practice.

Spaceball is a delightful low-gravity adaptation of the classic Earth game baseball. As you play, you'll discover a few key differences:

47

- Because it is played in a vacuum, pitches such as curve-balls, sliders, and knuckleballs are impossible because they rely on air pressure.
- A normal baseball on the Moon hit with the same force as a baseball on Earth could travel more than thirty times as far.
- Spaceball fields are larger than normal baseball fields, and the game can be dangerous because the balls travel so quickly (no air means no air resistance). Home runs aren't allowed and large nets collect stray balls.
- Though it's hard for players to run after a hit, once you get going it's more difficult to stop, which makes having quality cleats a must.
- Spaceball uniforms need to be stronger than average space suits in order to withstand accidental impacts.

Try a Lava Tube Excursion

Cavers will enjoy exploring the Moon's many underground lava tubes, ancient channels formed by flowing lava. They provide radiation protection from the sun and make ideal natural scaffolding for residences and buildings. While you needn't worry about encountering hot lava, the tubes are treacherous, with sharp terrain. You'll need a rover with long legs instead of wheels to navigate the rocky ground. If you decide to explore one of the many uncharted tubes, be sure to have a geologist check to make sure it is stable before venturing in; in rare cases they can hold early lunar atmosphere, which can be explosively released and knock you off your feet. The largest known cave is in the Sea of Tranquility.

Observe the Stars on the Far Side

During the Moon's two-week-long night, you'll relish uninterrupted stargazing. The Moon (and everywhere else within our solar system, for that matter) is close enough to Earth that you'll recognize the constellations. On Earth stars twinkle because the atmosphere bends the path of the starlight slightly this way and that as it travels through, but because the Moon has no atmosphere, stars don't twinkle there. Observers in the southern hemisphere can see the stars rotate around the lunar South Star, also known as Delta Doradus. Be sure to choose a helmet faceplate with an antireflective coating. This will help to cut down on stray reflections from lights on the surface, vehicles, or your suit.

The far side of the Moon is the ideal place for radio astronomy, which uses very long light waves to help astronomers reveal hidden

Giordano Bruno crater, one of the many craters on the Moon ripe for exploration

NASA/GSFC/ARIZONA STATE UNIVERSITY

views of celestial objects. The rings of Jupiter, which are much less obvious than the rings of Saturn, are more visible in radio waves compared to visible light. The radio telescopes are also sensitive to light from faraway galaxies, which has taken billions of years to reach you. The waves are the same as those used by microwave ovens and radios, but they were created by stars and gas in the distant universe instead of a small box in your house. The small craters of the lunar surface provide the perfect home for a large array of dishes used to detect the light. The far side of the Moon is removed from Earth's meddling atmosphere and electronic noise from its billions of residents, which is important because a single cell phone nearby can drown out the faint signal from a universe of galaxies. While you're visiting the far side, stop by Giordano Bruno crater, a fourteen-mile-wide pit with steep walls you may wish to try sledding down.

Tour a Moon Mine

Water is crucial for maintaining life on the Moon as well as for manufacturing rocket propellant. Don your hard hat and explore the subterranean industry on the Moon. Lunar rocks are rich in minerals such as plagioclase and anorthosite, which can be extracted to make aluminum. Ilmenite can be used for titanium and iron, and silica and oxygen can be extracted for use in materials engineering, manufacturing air, and rocket propellant. One of the most valuable natural resources is helium-3, an essential ingredient for fusion that is extremely rare on Earth but plentiful on the Moon. The far side is known for its helium-3 stores. High concentrations of it have been embedded in the regolith by a breeze of particles from the solar wind.

What's Nearby

After you've taken in the sights of the Moon, you can return to Earth or continue your journey to other planets. Just beyond the orbit of the Moon is a nice little haven called Earth-Moon L2, which is another Lagrange point. While you would expect a satellite that orbits farther from the Earth than the Moon to travel more slowly, at L2, the gravity of the Moon provides extra pull, allowing a satellite there to track with the Moon's orbit around Earth. Nestled here, you are far enough away from the pull of Earth's gravity to begin preparations for the next phase of your journey.

Destination ☿ Mercury

If you're looking for a sun-filled holiday, there is no better place than Mercury. A visit to this sweltering, moonless planet—the first rock from our sun—will be burned into your memory for years to come. To the average vacationer, the planet looks a lot like Earth's moon—a heavily cratered, mostly airless world only five hundred miles wider than our own Moon, with twice the gravity. But when you tour Mercury, you'll realize things are different here. The planet's surface is continually singed and sculpted by severe swings in temperature. The sun shines hot, but when it goes down the cold sets in.

Mercury attracts an audacious sort—sun worshippers unafraid of the imminent risk of death due to exposure. Whether you dare to flirt with the dangerous daytime climate of the surface or prefer the quiet cool confines of underground habitats hidden from the sun, your journey to Mercury will be filled with delightful contradictions. On this hottest of planets, there is plenty of shade, and even water ice, at its poles. With help from the limitless solar energy, provided by the nearby sun, you have a halfway decent chance of returning from your vacation to Mercury alive. Still, only the most courageous and determined adventurers decide to take the chance. It's the type of place you would brag about visiting, if you were the bragging type. Just remember, the accomplishment of surviving a Mercury vacation, no matter what you do while you're there, speaks for itself.

☿ AT A GLANCE

DIAMETER: 33 percent of Earth's

MASS: 6 percent of Earth's

COLOR: Gray with high-contrast shadows

SPEED AROUND THE SUN: About 106,000 miles per hour

GRAVITATIONAL PULL: A 150-pound person weighs
57 pounds

AIR QUALITY: Very little air to speak of; tiny amounts of
hydrogen, helium, and oxygen

MADE OUT OF: Rock

RINGS: None

MOONS: None

TEMPERATURE (HIGH, LOW, AVERAGE): 800, -290,
333 degrees Fahrenheit

DAY LENGTH: 4,222 hours and 36 minutes

YEAR LENGTH: 88 Earth days

AVERAGE DISTANCE FROM THE SUN: 36 million miles

DISTANCE FROM EARTH: 48 million to 138 million miles

TRAVEL TIME: 147 days for flyby

TEXT MESSAGE TO EARTH: 4 to 12 minutes

SEASONS: None

WEATHER: None

SUNSHINE: 5–10 times as bright as on Earth

UNIQUE FEATURES: Double sunset, the hollows

GOOD FOR: Sun worshippers, underground living

Weather and Climate

Mercury is a land of fire and ice. On its craggy exterior at midday, the sun shines nearly seven times as intensely as on the hottest day in the hottest desert on Earth. But after sunset, it gets colder than Neptune, and there are places on Mercury where the sun hasn't shone for billions of years. These dark places exist at the bottom of deep pits in the far reaches of Mercury's remote poles, protecting frozen water from the sun's scorching light.

If you could handle the heat on the surface, you'd experience thousand-degree temperature swings between night and day. At dawn, it's a cool -290 degrees Fahrenheit. As the sun rises higher in the sky, the ground starts to warm up, eventually topping 800 degrees. There is no atmosphere to transmit the heat, so the rays from the sun would bake you directly, and heat would seep up from the hot ground, if you went outside at noon. At night, when you'll be able to visit the surface, you'll need a well-heated, insulated space suit to prevent you from radiating away your precious warmth. Make sure you have insulated boots, or your feet will freeze solid.

The days are long and the years are short on Mercury. It zips around the sun more quickly and has a shorter distance to go in its orbit than Earth, so its year lasts just eighty-eight Earth days. The sun's uneven pull between the near and far sides of the planet has slowed Mercury's spin, and so the planet rotates just once around its axis every fifty-nine Earth days. But Mercury's *solar* day—the time it takes for the sun to appear at the same point in the sky—lasts 176 Earth days, equivalent to three planetary rotations and two Mercurian years. That means the solar day on Mercury is longer than its year.

What we know as winter, spring, summer, and fall are unknown on this alien planet. The tilt we have on Earth that helps determine the seasons does not exist on Mercury, so the poles never nod toward or away from the sun. Our Earth-oriented understanding of the four seasons is replaced by the binary environmental conditions of acute hot and cold.

You don't need to worry about storms while wandering out and about. The planet has no actual wind. The only breeze is the constant flow of high-energy particles from the sun, called the solar wind.

When to Go

In the dead of winter, when you've had enough doubling and tripling up of blankets and you don't think you can shovel another pound of snow, think of the sun toasting Mercury's dusty craters. When you need a little sunshine in your life, Mercury delivers. In just twenty-four hours on the planet, you'll get a whole Earth week's worth of sunlight.

There is such a thing as too much sun, however, and on Mercury it's easy to overdose. The sun has its own pattern of storms, and every eleven years it gets feisty and has an increase in high-energy flares. On Earth, this solar weather interacts with our magnetic field and can disrupt electronics. On Mercury, a solar storm can abruptly end your holiday, so check the space weather report before you go.

Mercury has an elliptical orbit, and its hot season occurs when it's closest to the sun. It's wise to avoid visiting at this time of year. If you visit when it's farthest from the sun, the maximum daytime temperature drops to a mere 530 degrees from its high of 800. You'll also be farther from any unexpected solar storms.

Because Mercury has such a short year, it doesn't take too much advance planning to time your trip to a particular point in the year. You can experience the breadth of the intense hot and cold conditions on Mercury in only six Earth months—perfect for mercurial types.

Getting There

Because Mercury moves so fast, getting there is trickier than you might think. Tickets can be pricey, though there's no world in the solar system where the sun shines brighter. It's at least 48 million miles away and because of its speed around the sun—106,000 miles per hour, about 40,000 miles per hour faster than Earth's—it takes more fuel to leave the solar system than to speed up to meet pace with fleet-footed Mercury without accidentally getting pulled into the sun. Some of the more fuel-efficient flights can take up to eleven years, as your ship completes a series of gravity assists from Venus and Earth that will eventually place you into orbit around Mercury. It's possible to get there in as little as 147 days, but only if you're willing to whiz past without stopping.

Solar energy is plentiful on Mercury, and eco-minded vacationers appreciate that it's a renewable resource—at least for the next 10 billion years, until the sun dies. Keep an eye on how much juice your solar panels are producing as you approach the sun to prevent your electronics from frying. If you're getting too much energy, you can turn them to face away from the sun.

As wonderful as all that sunlight is, the sun's gravity is a cause of annoyance for navigators because ships tend to get tugged in its direction, like a car that needs its wheels aligned.

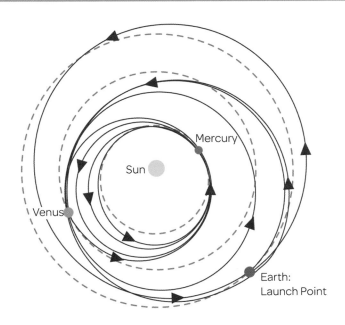

Though Mercury is closer than many other planets, getting there is made complicated by the planet's great speed and proximity to the sun.

When You Arrive

As your ship approaches Mercury, you may wonder if you're actually visiting Earth's moon. However, once you land, you'll see that the rock of Mercury is different in hue and texture. You'll notice contrast—after your eyes adjust to the blackness of the sky, you'll discover that brighter things look brighter and darker things look darker.

The surface—the shade of pencil lead—is darker than the

Moon's because of the presence of graphite. Although it has a huge molten iron core, its surface lacks iron. It's mostly graphite gray with a hint of reddish brown, and some areas appear to have a bluish tint. Lava channels resembling dry rivers and mysterious hollows—areas of the surface distinguished by their ever-shifting appearance—hint at the active geology of a changing planet.

You'll be impressed by the variety of terrain across Mercury's landscape. It has weathered violent collisions, volcanic eruptions, and planet-level contractions that have ripped apart its surface. It's home to huge cliffs, double-ringed craters, trenches, and mysterious hot and cold spots.

Though impacts from above have shaped much of Mercury's surface, the planet also has some spectacular tectonic forces that have molded its face—cracking it open to form valleys, squeezing it to form long sinuous cliffs called lobate scarps, or wrinkle ridges. These cliffs line the surface for hundreds of miles.

You'll land at night, and before the sun rises you'll need to descend from the dusty surface into an underground city. Even the temperature underground would be far too hot to survive during the day if it wasn't for the highly reflective roofs that, over the course of a decade or so, deflect enough light for the ground beneath to cool to a survivable temperature.

If you're longing to see the daytime sky, you can view it through a periscope while safely staying deep below the surface. You'll see a vast blackness punctuated with pinpoints of starlight. Though technically there is the slightest hint of an atmosphere on Mercury, it's not nearly enough to make a sky, or enough to make the stars twinkle. Keep searching and you'll find the unusually brilliant sun, sharing the dark sky with the stars.

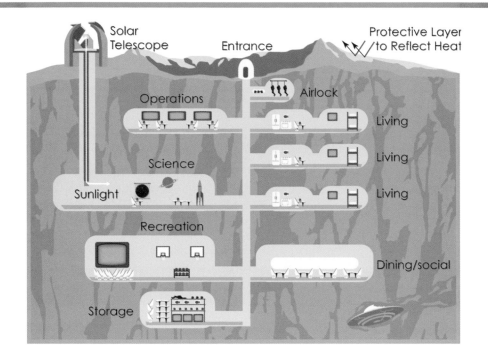

Getting Around

Though Mercury is not the hottest planet in the solar system (that honor goes to Venus), it's hot enough to make certain aspects of Mercurian life complex. Your trip to the surface will require careful planning because at any given time, half of the planet is bombarded with raw radiation from the sun, unfiltered by an atmosphere, making it uninhabitable.

Some prefer to stay in the relatively stable but cold comfort of the permanent darkness offered by some of the craters at the planet's poles. Those who venture outside the poles must acclimate to dwelling underground for several Earth months at a time, when surface activities are banned due to the brutal temperatures. Sunrise brings a sense of dread to the underground dwellers, who know if you're caught outside at the wrong time, you'll meet a blistering death from heat.

You can rent rovers or hoppers for regional night tours. For longer treks, you'll book a seat on an intraplanetary rocket. A delayed launch can lead to an unexpectedly long layover as you wait for surface travel to resume. Have patience. You might spend the forced periods of subterranean sequestration studying the work of

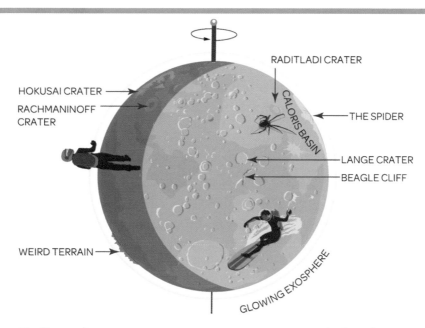

HOKUSAI CRATER
RACHMANINOFF CRATER
RADITLADI CRATER
CALORIS BASIN
THE SPIDER
LANGE CRATER
BEAGLE CLIFF
WEIRD TERRAIN
GLOWING EXOSPHERE

You'll get to know Mercury's many secrets in its protective shadows facing away from the sun's dangerous rays.

the famous artists for which many of the planet's landmarks have been named. During your wait, take a two-month dive into Chekhov's letters, watch the complete filmography of silent film star Billie Burke, or study the ways of Japanese tea ceremony master Rikyu.

What to See

The North Pole

As you approach the north pole from orbit you'll see a stunning view of the sixty-two-mile-wide Hokusai crater, named for the eighteenth-century Japanese artist Katsushika Hokusai, who's best known for his wood-block print of an enormous wave. The crater's long spokelike rays span thousands of miles from the center, formed by rock ejected from a violent asteroid impact. Covering most of Mercury's northern hemisphere, it has some of the biggest rays in the solar system.

The north pole maintains a nice, cool year-round temperature of -136 degrees Fahrenheit. It's close to the coldest temperature ever recorded on Earth. Both of Mercury's poles contain water ice, formed in shaded craters that never get sun. In the north, the ice sheet is tens of feet thick and ten thousand times smaller than the Antarctic ice sheet. Better than nothing!

Craters forever cast in shadow have stable, cold environments and are a draw for scientists looking to excavate billion-year-old ice cores to study the impact history of the solar system. One theory they're testing is whether an ice-filled comet was responsible for creating the Hokusai crater, showering the poles with ice knocked loose during the impact.

Bright rays lead you to Mercury's impressive Hokusai crater, seen here near the edge of the image.

Ice climbers will enjoy scaling the rims of craters in the region. Though there are a few reasonably steep slopes, skiing is made challenging by the cold temperatures, which keep the ice as hard as rock.

Prokofiev crater—named for the early twentieth-century Russian composer of *Peter and the Wolf*, Sergei Prokofiev—is the largest crater in the north pole. The depression has large regions of ice, which scientists believe fell to the ground long after the crater was formed.

If you have a few weeks to spare, you'll want to explore the north pole's vast volcanic plains south of the cluster of craters near the center of the pole. You will never feel smaller than when you are standing in the middle of a lava field hundreds of miles wide, the product of an unimaginably powerful eruption that covered the area in glowing hot liquid billions of years before.

The Caloris Basin

Once you've sufficiently cooled off at the north pole, it's time to heat things up at the Caloris (Latin for *hot*) basin. It's a short trip, and on your way you'll enjoy the aerial view of five broad valleys named for abandoned cities of antiquity: Cahokia, Caral, Paestum, Timgad, and Angkor. As you continue south, look carefully to the east and you might be able to get a peek at Oskison crater. It's named for American Indian writer John Milton Oskison and serves as a destination for authors who want to take up peaceful residence within the peaks at the center of the seventy-five-mile-wide crater.

You can land in Atget crater, just south of the center of the basin. The dark-floored sixty-two-mile-wide depression is named for Eugène Atget, a French photographer known for his extensive documentation of late nineteenth- and early twentieth-century Paris. From Atget you can explore the immense smooth lava plains of the Caloris basin. One of the largest impact craters in the solar system, it stretches over an area the size of Alaska, more than nine hundred miles across and surrounded by a ring of mountains over ten thousand feet tall. Beyond the main mountain ring are hundreds of miles of rough, hilly plains and lines caused by land thrown out by a huge impact. Experts believe the enormous space rock that carved out the basin billions of years ago was dozens of miles wide, and that its collision with the planet was so violent it created bumps on the opposite side of the planet. They're known as the "weird terrain." The lava flows that created the plains would have covered a huge area and been far less viscous than what flows from volcanoes on Earth.

There is plenty to explore within Caloris, including craters named after gothic American author Edgar Allan Poe and Norwegian painter Edvard Munch, of *The Scream* fame. But the most intriguing attraction is the Pantheon Fossae, a series of trenches nicknamed "the Spider" for its resemblance to a creepy crawler. The formation has a central twenty-five-mile-wide crater surrounded by more than a hundred narrow canyons, called graben, expanding outward. Each is several miles wide and some stretch for more than two hundred miles. No one is exactly sure how these valleys formed, and whether Apollodorus crater at the center—named for the architect of the Pantheon—had anything to do with it. As you hike along the valleys, you can look for clues to solve the mystery of the region's enigmatic past. The formation looks unlike anything else on the planet and defies conventional understanding of planetary geology.

You can take a day, a week, or a whole month exploring the trenches. For those inclined to take the easy path, you can catch a shuttle between Atget crater and Apollodorus in the center. Just remember all shuttle service, and surface activity, will cease when the sun comes out. You don't want to get stuck in one of the valleys near morning.

When the sun finally goes down again, you can continue your exploration of Caloris with a visit to Cunningham crater, named for American photographer Imogen Cunningham. Mercury's high-contrast lighting provides an ideal backdrop with which to perfect your black-and-white photography skills. West of the Caloris basin, you can stop for a night at Kerouac crater. The sixty-eight-

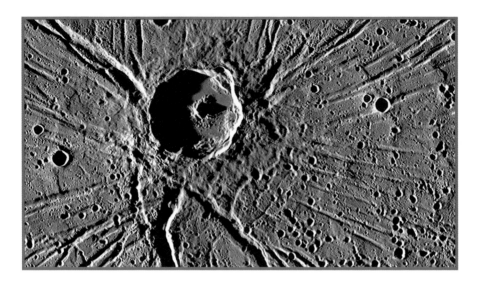

See Apollodorus crater and the Pantheon Fossae, also known as "the Spider," when you visit the Caloris basin.

NASA/JOHNS HOPKINS UNIVERSITY APPLIED PHYSICS LABORATORY/CARNEGIE INSTITUTION OF WASHINGTON

mile-wide depression is an enclave for the planet's alternative crowd. Take a pilgrimage to the center of the crater and wax poetic as your Earth-centered view of the universe wanes.

The Hollows at the Raditladi Basin

One hundred and fifty miles west of Kerouac crater is the Raditladi basin. You'll stand at the center of the young (only 1 billion years old), 160-mile-wide crater and notice two ridges in the distance. One is thirty-nine miles away; the other, eighty miles. These are the "double rings" of the crater. The basin also has troughs like those found in the Pantheon Fossae. They were possibly formed during an "extension," or large-scale stretching and contracting of Mercury's surface. Rather than radiate outward, these shallow valleys form concentric rings that ripple from a point six miles from the center of the basin.

The most intriguing mystery of Raditladi, named for Botswanan playwright and poet Leetile Disang Raditladi, is its sulfur hollows. The patterns in the hollows are recently formed, showing that Mercury is still changing, with a surface that seems to be alive with activity such as hot springs. They are very reflective and look bluish compared with the rusty gray of the rest of Mercury. Found on other parts of the planet's surface, these funny-looking depressions might form when heat evaporates material in the ground. At night they are fairly stable, but during the day some of the material is thought to turn straight to gas, a process known as sublimation.

The hollows are easy to climb, ranging in depth from thirty feet to the height of a ten-story building. They are up to a mile long. When you're standing inside one, you'll feel like you're in a small canyon but with brighter walls. Just be careful as you traverse the hollows; the rock is riddled with holes and can unexpectedly crumble as the hollows gradually get wider. An experienced tourist might notice a different texture or crunch underfoot on this younger land, compared with other, older areas generated by impacts over eons. Other good places to view hollows include the craters Kertész and Zeami.

Beagle Cliff

To the south of Raditladi you can walk along Beagle Cliff, a nearly four-hundred-mile-long, one-mile-high cliff that cuts through several craters. Dozens of cracks in the land like this one formed in Mercury's early history, when the entire planet contracted as its molten iron interior cooled. The cliffs are named for the ships of famous explorers. Beagle Cliff is named for HMS *Beagle*, the ship that permitted Charles Darwin's extensive scientific observations of South America and Australia. To the northwest of the cliff is Lange crater, after early American photographer Dorothea Lange, famous for her 1936 photograph *Migrant Mother*.

Rachmaninoff Crater

The Rachmaninoff crater, a 180-mile-wide double-ringed basin, is one of Mercury's youngest. The eighty-mile central ring contains smooth reddish plains thought to be left from volcanic lava flows. The southern part of the ring appears to have been immersed in these flows. If you're lucky, you'll catch a silent Rachmaninoff night concert under the stars. An orchestra won't make any sound because there is no air for it to move through. Think of it as Mercury's salute to John Cage.

Activities

Double Birthday

On the best vacations, time seems to stop and you wish your perfect day could last forever. Vacationing on Mercury may be the closest you'll come to realizing that dream. During Mercury's 4,224-Earth-hour (176 Earth days) solar day, you can pack a lot in. It also means you celebrate your birthday twice per day, and you can live well into your three hundreds—as long as you're counting in Mercurian years.

Watch a Double Sunset

One of the greatest reasons to visit Mercury is to witness the peculiar behavior of the sun in its dark sky. Of course, unless you have a death wish, you won't be seeing the sun's dance across the sky firsthand. You'll be able to safely view—and celebrate—the sun's acrobatics through an underground telescope. From this unique vantage point, you can watch the sun as it takes its sweet time to pass overhead. There is a special moment each day when it appears to stop and reverse course temporarily, before continuing on its several-Earth-months-long journey to sunset. Depending on where you are on Mercury, the sun may even appear to set and then rise again, as the planet passes through the point of its closest approach to the sun and its angular speed is temporarily faster than its rotation. Take this moment to reflect on the past and think about what you might do if you could turn back time. Forgive old grudges and right any wrongs you've committed over the previous Mercurian morning (i.e., year).

When the sun is moving backward in the sky, you may notice that it looks gigantic. Don't worry, it's not sun poisoning affecting your judgment. Because of Mercury's closeness to the sun, the sun does appear larger in the sky than it does from Earth. The planet's highly elliptical orbit enhances this effect, and the sun will seem to grow and shrink throughout the year. At Mercury's closest approach, 29 million miles—known as perihelion—the sun looks three times as big as it does on Earth. During aphelion, when Mercury is at its farthest point, 43 million miles from the sun, it only looks about twice as big.

One of the most fascinating natural phenomena in the solar system is the sun moving backwards in Mercury's sky, a quirk of its fast speed around the sun and slow rotation.

Sail the Solar Wind

Unfurl your reflective sails on a solar sailboat. Photons, or light particles, streaming away from the sun will bounce off the mirror-like surface, giving your ship a miniscule push. The combined effect, called radiation pressure, will let you sail the seven seas of the solar system. You might not accelerate dramatically, but you will slowly gain speed over time until you're going fast enough to travel astronomical distances. If you get caught in a solar flare, you'll have to batten down the hatches.

Walk the Eternal Sunset

The primary reason to visit Mercury is the chance to walk its slow-moving terminator line, which divides day and night. Because the planet rotates so slowly, it's possible to stay ahead of the sunrise. The terminator on Mercury travels at a reasonable walking rate of just 2.2 miles per hour, while the line that divides day and night on Earth travels at around 1,000 miles per hour. The terminator provides dramatic shadows, and this stark safe zone between night and day maintains a surprisingly habitable climate, though the consequences of falling behind can be fatal. Don't get caught in direct sunlight, or you will be just another lump of human charcoal dotting the Mercurian landscape. Extreme athletes love the challenge of walking the entire 9,500-mile distance around the planet in one shot.

Mercury's terminator is not a cyborg from the future but rather the slowly moving line between the lit half and the dark half of the planet.

NASA/JOHNS HOPKINS UNIVERSITY
APPLIED PHYSICS LABORATORY/
CARNEGIE INSTITUTION OF
WASHINGTON

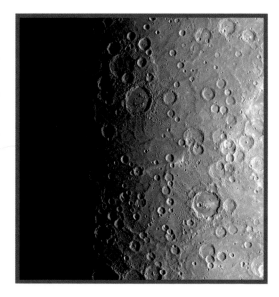

Ski the Volcanic Sands

The fine-grained sands left behind from billion-year-old volcanic eruptions are fantastic for sandboarding and skiing. Low gravity provides the perfect setup for getting major air. There are dozens of these so-called pyroclastic deposits all over the planet, including one in the southwest of the Caloris basin and one west of Copland crater.

Experience the Supernatural Glow of the Exosphere

Mercury's very tenuous atmosphere, its exosphere, is rich in sodium. When night falls, an amber-yellowish glow encases the sky, like sodium-vapor lights in a parking lot. It will remind you of the auroras back home on Earth, but it diffusely covers the whole sky and is brightest toward the horizon. The light is enough to see by, which is fortunate because all of your surface activity must occur at night.

Visit the Ghost Ships

Scientists lost touch with the first spacecraft to visit, *Mariner 10*, when it ran out of fuel back in the 1970s and drifted into space. No one has heard from it since. The ghost of *Mariner 10* haunts the inner solar system. Legend has it the spacecraft keeps calling to Mercury, still trying to study it, supposedly collecting data and eerily sending it into the void of outer space. You can also see the remnants of another dead ship, the *MESSENGER* craft, which intentionally crashed into the planet on April 30, 2015. The impact left a crater on the surface fifty-two feet wide.

Destination ♀ Venus

They say Venus is for lovers. And as long as you stick to the temperate climate high above the planet's scorching surface, Venus is true to its namesake, the Roman goddess of love, beauty, and desire, providing a relaxing atmosphere in which to enjoy your time away from the stresses of work. Many vacationers decide to visit after seeing it as a bright star beckoning them from Earth's night sky. It is a place well suited to introspective romantics.

Our nearest neighbor, Venus is often called Earth's hot twin. Though it's roughly the same size as Earth, with similar gravity, it's more than eight hundred degrees warmer, on average. As the hottest planet in the solar system, hotter even than Mercury, it's not surprising that many naive tourists believe the weather on Venus is simply intolerable. What most don't appreciate is that within its skies, Venus has one of the friendliest environments in the solar system, and there are plenty of diversions on the surface for those who take appropriate precautions.

The floating cities of Venus kiss divine cloudscapes, and you'll feel like you're in heaven, adrift among its sulfuric acid mists. Just make sure your air tank is full and you've donned your acid-proof clothing. If you don't get airsick very easily, enjoy gazing at clouds, and aren't bothered by high pressure surroundings, you may be ready for a rendezvous with Venus.

♀ AT A GLANCE

DIAMETER: Slightly smaller than Earth's

MASS: 81 percent of Earth's

COLOR: Golden to reddish-brown bathed in yellow light

SPEED AROUND THE SUN: About 78,000 miles per hour

GRAVITATIONAL PULL: A 150-pound person weighs 136 pounds

AIR QUALITY: Thick, 96.5 percent carbon dioxide and 3.5 percent nitrogen

MADE OUT OF: Rock

RINGS: None

MOONS: None

TEMPERATURE (HIGH, LOW, AVERAGE): 867, 867, 867 degrees Fahrenheit

DAY LENGTH: 2,802 hours

YEAR LENGTH: About 7.5 Earth months

AVERAGE DISTANCE FROM THE SUN: 67 million miles

DISTANCE FROM EARTH: 24 million to 162 million miles

TRAVEL TIME: 100 Earth days for flyby

TEXT MESSAGE TO EARTH: 2 to 15 minutes

SEASONS: Very mild

WEATHER: Slow but strong winds; acid rain

SUNSHINE: Almost twice the brightness of Earth's

UNIQUE FEATURES: Floating cities

GOOD FOR: Heat seekers, daydreamers

Weather and Climate

The most Earth-like climate in the solar system (outside of Earth) lies thirty-four miles above the surface of Venus. Like the famous floating city of Laputa, from *Gulliver's Travels*, it's the perfect distance to forget about the brutal conditions below. The temperature here is balmy but perfectly manageable, in the nineties Fahrenheit, and the pressure is similar to that on the surface of Earth.

The surface is a different story, however. Unless you have a taste for the supertropical, Venus's sweltering soil is a nightmarish inferno. This is what the hottest planet in the solar system feels like. Ninety-six percent of the air is made up of carbon dioxide, the heat-trapping greenhouse gas responsible for warming on Earth. The thick smog traps so much heat that the temperature reaches a searing 867 degrees. If there were oxygen in the atmosphere, a stray piece of paper would spontaneously burst into flame. Though Venus is closer to the sun than Earth, without the greenhouse effect to hold in heat, the surface would be a chilly 8 degrees.

The solid terrain of Venus is hellish, to be sure, but it's very predictable. Buffered by the thick blanket of atmosphere, the temperature doesn't swing much from day to night, or over the course of a Venusian year. Seasons are virtually nonexistent, as the tilt is low and the sun's rays are distributed more or less evenly. The year on Venus lasts just over seven Earth months, and the time it takes to rotate once around is the longest of any planet in the solar system—243 Earth days. This takes longer than the local year, which lasts 225 Earth days. The rotation is so slow that it's also longer than the planet's 117-Earth-day-long solar day, which is the time it takes for the sun to move from one noon to the next.

Venus has a strange backward or retrograde motion, rotating in the opposite direction of all the other planets except Uranus. If you could see the sun through the clouds from the surface and watch it for a whole Venusian day (good luck staying awake), you'd see the sun rise in the west and set in the east. The reason for this peculiar behavior is a mystery, though it may be the result of a huge asteroid collision long in the past.

The clouds of Venus in real color, with slightly more detail than your eye will see

MATTIAS MALMER/NASA

Any trace of water from Venus's ancient oceans has long since evaporated. If you suddenly poured an ocean of water on the surface, it would turn bubbly like soda water before evaporating away. The high pressure, carbon dioxide–rich atmosphere of the planet works like a giant soda carbonator.

A very slow, constant breeze rolls over the surface. This wind can be surprisingly powerful, because the thick air easily blows over light objects. Once in a while the weather turns bad, with volcanic eruptions triggering widespread acid rain, which evaporates before making it to the surface. Venus also has lightning, similar to that on Earth, striking between the sulfuric acid clouds. Look for an eerie flash in the orange-yellow skies—if you can see through the haze.

When to Go

Anytime is a good time to visit Venus, as long as you stick to the airships in its temperate climate high above the surface. It's a reliable sanctuary for anyone eager to do some soul-searching. The trip is long, but not too long. You'll be gone for several years—enough time to completely transform your perspective. Recent college graduates can take time off to go and contemplate their existence. It's also a good place to get away to rejuvenate and reset after a major life transition.

Getting There

Remember that your future vacation spot is a moving target. Venus is flying around the sun at 78,300 miles per hour, almost 12,000 miles per hour faster than Earth is. It's 23.7 million miles away at its closest, and you'll need to step on the gas to catch up to it. Bridging the speed gap by brute force—simply firing your rocket—can alter your orbit in unexpected ways in addition to using a lot of fuel. Remember: Fuel is weight, and weight is money.

One way to travel between two planets is called a Hohmann transfer orbit, a low-energy path that connects two orbiting objects. It's named for Walter Hohmann, the German scientist who invented it in the early twentieth century. If you choose this flight plan, you'll travel in an elliptical orbit that touches both planet's paths; you'll arrive in about five months. Since Earth and Venus are both in motion, you only have one chance—a launch window—every nineteen months to fly along a Hohmann transfer orbit. Try not to schedule your flight too close to the end of a launch window, because if your launch is scrubbed you'll have to wait a year and a half to try again.

You might think that starting from Earth you'd need to speed up your spacecraft to catch up to Venus. However, rocket science isn't so straightforward. To match orbital paths and speed with Venus, you'll need to fire your thrusters in the opposite direction of Earth's motion around the sun. This will bring your orbit closer to the sun and you'll naturally end up going faster.

The good news about getting to Venus is that it has a thick atmosphere, which doubles as a cheap braking system. When you arrive you can surf along the top of it to slow down. Use caution; it's easy to accidentally burn up in the Venusian air because it's so thick.

When You Arrive

After a long stretch in the blackness of space, you'll be blinded by the brightness of Venus. It's not the surface that is bright, but rather the

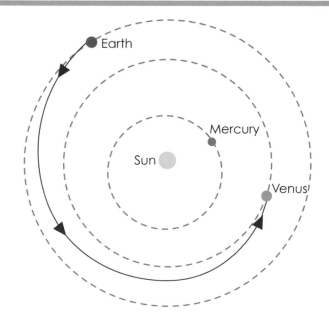

Neighboring Venus is one of the hottest planetary holiday spots.

ethereal Venusian cloud tops. Bobbing in and out of the thick cloud deck are the glittering cities in the sky. These havens of safety and luxury will be the first stop on your tour.

While the skies look familiar and inviting, the carbon dioxide–rich air of Venus can quickly asphyxiate you. The upside is that Earth air floats easily in the thick sea of Venus's gases. Habitats filled with breathable air can rise like a helium balloon does on Earth. Tensegrity City, named for the airborne geodesic habitats first conceived by the ingenious twentieth-century inventor Buckminster Fuller, is as big as an Earth city. It is sealed off from the air surrounding it, and filled with Earthlike air that keeps it aloft in the Venusian sky. When you look out the windows you'll see you're nestled in a thick layer of puffy white clouds. Don't be taken by their cheerful countenance; many are composed of corrosive sulfuric acid.

Although the day is 2,802 hours long on the ground, the floating cities in the sky move with the clouds, which travel around the planet in just one hundred hours. That means you have to adjust to fifty hours of daylight and fifty hours of night. The local custom is to sleep once within a twenty-five-hour period. Two of these sleep cycles will be in daylight, while two will be in darkness. A hotel room with large windows for cloud viewing and good blackout shades will help you sleep when the sun is out.

Since you'll be wobbly on your feet for a while thanks to the winds, plan your holiday so there's enough time to get your sky legs. Some visitors get them immediately, while others might feel queasy for a week or so. In addition to the gentle rocking, turbulence is a perpetual threat for sky dwellers. Like storm warnings on Earth, turbulence alerts will let you know it's time to find a nearby safety shelter where you can hunker down through the rocky weather.

If you choose to visit the surface, you'll notice a drastically different setting. On the ground, Venus is both an oven and a pressure cooker. The pressure at the surface is ninety times the pressure at sea level on Earth, similar to what you'd encounter diving three thousand feet below the ocean.

What you see there might just be worth the trouble, though. Deep below the cloud deck, the light on Venus has a lovely orange haze. Blue light scatters more easily than red light when it travels through the air, which is why Earth's sky is blue. In the extra-thick air of Venus, this scattered blue light is absorbed more strongly. Only the orange light remains, leaving Venus's sky bathed in a perpetual apricot hue. It might remind you of sunsets on Earth—except you can't see the sun because the clouds are too thick. Some say the low light levels enhance the eerie ambiance. In this dusky scene, you may see visual distortions. Much like a desert mirage, light from the sky or distant objects may be bent. This makes navigation a nuisance, since you can't trust your eyes to give you accurate information about your surroundings. Instead of trying to eyeball how far away objects are, use maps or satellite navigation to orient yourself. Adding to the disorienting scene are the strange sounds of Venus. In the thick, soupy atmosphere, everything sounds much deeper than usual, rich in bass, and distorted into a spooky roar.

Getting Around

There are many ways to navigate the clouds of Venus. The air is thick and great for gliding. Light Venusian airplanes bear a striking resemblance to airplanes on Earth. They take advantage of Venus's proximity to the sun by running exclusively on solar energy. As long as you stay above the top layer of clouds, where the sunlight is twice as bright as it is on Earth, there is abundant power for a light aircraft. Be sure not to accidentally dive into the clouds or fly too far into the night side or you could lose power and plunge to a blistering death.

If the crowds of Tensegrity City are making you antsy and you want an escape, reserve a personal airship and disappear into the thickness. Just large enough for one or, if you're feeling very cozy, two, airships launch from the city and you can catch an air

Cruise the cracked turf of Venus in a hardy vehicle.

current for a tour of the clouds. Be careful not to get blown off course by jet streams.

For those craving a closer look at the legendary lava plains below, be warned that flying down to the surface is like diving thousands of feet into Earth's ocean, except it's thirty times hotter. Under those conditions, you'll need a vehicle that looks like a submarine on wheels. This hardened RV will be your home as you cruise the arid landscape. You'll need an air-conditioning system that can withstand the hellish weather.

What to See

Ishtar (Ishtar Terra)

Named after the Babylonian goddess of love, Ishtar is an island surrounded by dry land. Compared to continents on Earth, Ishtar is a bit larger than Australia. On its eastern ridge are the Maxwell Mountains, whose peaks are taller than Mount Everest. It also contains the volcanoes Sacajawea and Cleopatra. Within Ishtar is the Lakshmi plain (Lakshmi Planum), which is similar in height to the Tibetan plateau of Earth. In Babylonian mythology, Ishtar travels to the un-

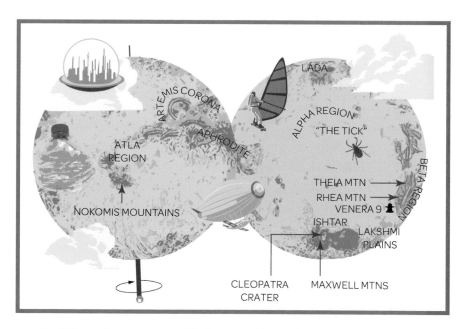

You'll have plenty to see and do by air and by land.

derworld, where the goddess of the underworld makes her leave a piece of clothing at each of the many gates. If you dare explore Ishtar, keep your clothes on. Without your space suit, you'll lose consciousness after a few breaths and suffocate shortly after.

Aphrodite (Aphrodite Terra)

As large as Africa, this raised area just skirts the equator and extends into the southern hemisphere of Venus. Visit the dual plateaus of the Ovda region (Ovda Regio) and Thetis region (Thetis Regio). Giant earthquakes and tectonic pressure have heaved the land against itself, creating huge tilelike deformations in the ground. The ridges and cliffs make for a strange, crisscrossed landscape.

Alpha Region (Alpha Regio)

The pancake dome chain, just east of Alpha region, is a set of seven overlapping circular hills that look like giant flapjacks. They are round, flat, and large, with cracks in their surface. Pancake domes are a unique type of volcano not found on Earth, thought to form when thick lava erupts from the surface, cools, and vents gas. Keep a special eye out for "the Tick," a volcano more than one hundred miles across with ridges and valleys that look like the legs of a tick when viewed from above.

Beta Region (Beta Regio)

This highland has two large mountains, Theia Mountain (Theia Mons) and Rhea Mountain (Rhea Mons). While they look like twin moun-

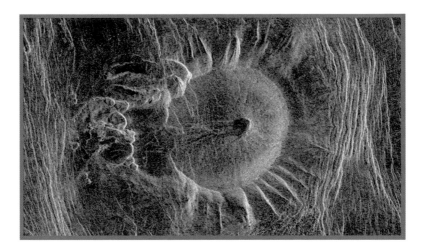

Visit "the Tick" on Venus, a volcano named for its unique shape. The body is 22 miles wide.

tains, Theia is a volcano, while Rhea is not. Theia is a shield volcano—a feature named for its wide, gentle slope—like Mauna Kea on Earth, built up with layer after layer of fluid lava over many years. South of Rhea Mountain and Beta region, Devana canyon (Devana Chasma) is one hundred miles across and four miles deep at its lowest point.

Lada (Lada Terra)

Flanked by Helen Planitia, Lavinia Planitia, and Aino Planitia, this raised area—a landmass with no sea surrounding it—has some lovely landscapes. Visit the deep rift valleys, viewing the depths from the rims. Be sure to visit the Quetzalpetlatl corona, a ring formed by volcanism and tectonics.

One of the first photos ever taken of the Venerean surface

Early Probe Landing Sites

Take a tour of the remains of dead robots. In 1961, the Soviets started sending probes to Venus to study it. None survived long in the punishing conditions, and the first ones perished before they could send any useful data back. Engineers learned through an expensive process of trial and error how to build one that could tolerate the hot air. By the eighth try, the probes could last for an hour before dying. *Venera 9* and *10* provided our first pictures of the surface. NASA's *Pioneer Venus 2*, launched in 1978, transmitted very briefly from the surface after probing Venus's atmosphere. Many enjoy visiting the crash sites, but brace yourself for robotic carnage.

Activities

Take a Stroll Through the Clouds

Viewing decks allow you to get some time outside the controlled environment of the city, though you won't be getting any fresh Venusian air. The pressure at the altitude of the floating cities is similar to Earth's, and as long as your skin is fully protected from the acid mists and you have a source of breathable air, you can enjoy the cloud tops of Venus without a bulky pressurized suit to interfere. Always check the weather forecasts and city altitude to make sure you won't be traveling through thick sulfuric acid clouds or high winds. Severe acid burns are sure to wreck a perfectly pleasant outdoor walk.

Stargaze

See Earth from above the clouds during Venus's long nights. With the occasional bumps and jolts, sky watching from an airship has its drawbacks. Your telescope requires intricate stabilizing mounts for smooth viewing. Binoculars work well in a pinch, and with them Earth appears as a bright, slightly bluish dot. Mercury, being closer to Venus, is easier to observe from Venus than from Earth, while Mars is dimmer. You should also be able to see the Moon. The rest of the sky will be very familiar to stargazers from Earth.

Terrasurf

Surfing the terrain of Venus is a lot like windsurfing on Earth, minus the cool breezes and sparkling oceans. The air is thick, and as you glide along the ground on a covered board with wheels, you can easily catch a slow breeze to pull you along the dusty land. Your wheels need to be large and tough to weather the sharp rocks. With good technique you can tack into the wind and terrasail faster than the local wind speed, which is just a few miles per hour. Atalanta plains (Atalanta Planitia), one of Venus's largest basins, is a fabulous place to bring your surfboard. Its smooth terrain makes it ideal for racing.

Crush It

While it may not be the most glamorous sport, many visitors to the cloud cities of Venus get a kick out of crushing plastic bottles. The first step is to cast a very long string down toward the surface with a sealed bottle on the end, much like casting a long fishing line. If you lower it far enough, you'll get to an area with higher pressure, which will crush the bottle into a crumpled mess. Lower it too far and your plastic bottles will melt entirely.

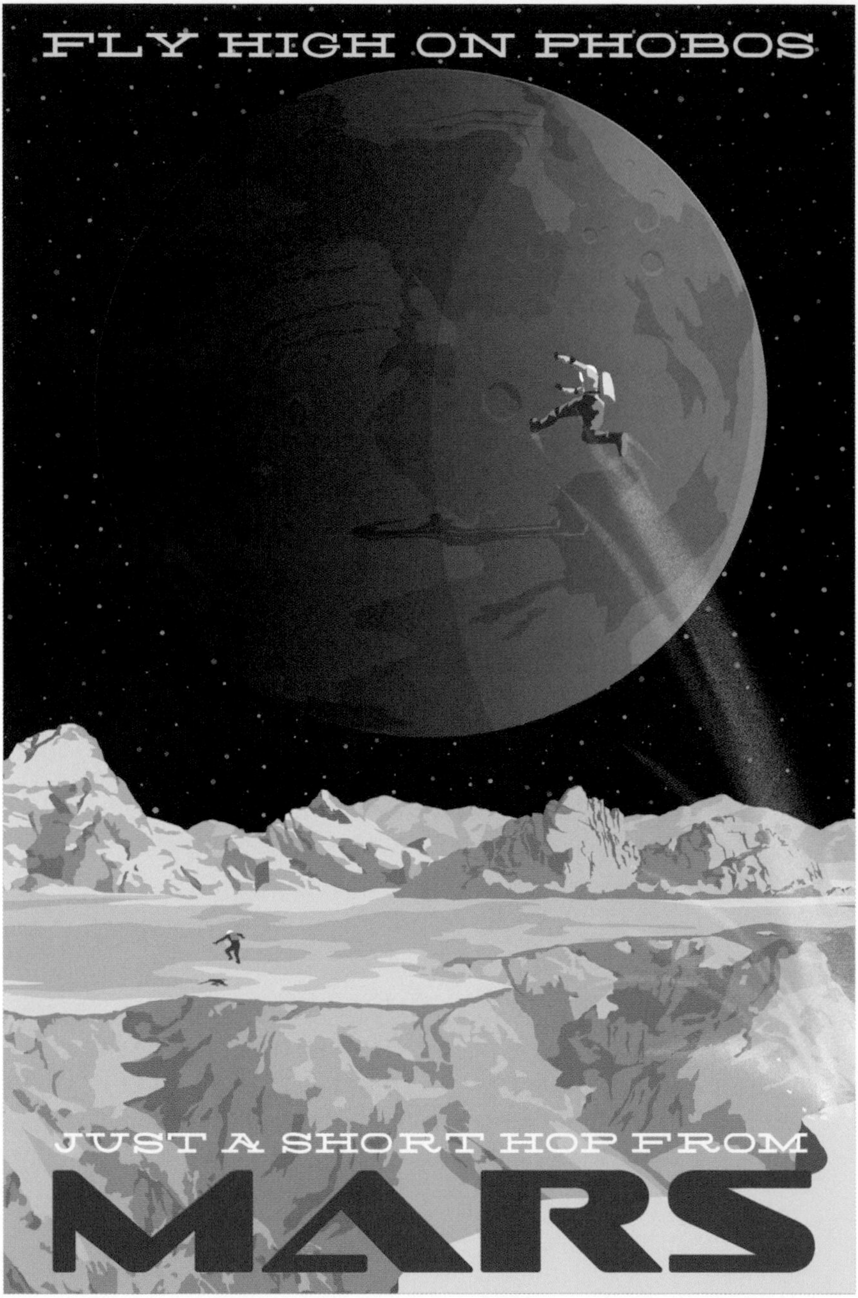
FLY HIGH ON PHOBOS
JUST A SHORT HOP FROM
MARS

Destination ♂ Mars

Who hasn't dreamt of setting foot on the red planet? With butterscotch skies, colossal canyons, and the tallest volcano in the solar system, Mars is an oasis for romantics and adventurers alike. The vast, frigid deserts are exotic, yet strangely familiar. It's like a smaller version of Earth in some apocalyptic parallel universe, where the oceans have dried up and the atmosphere has drifted away to almost nothing. What's left is dust, rock, and a plethora of luxury resorts where you can escape the weight of all that gravity and heavy air of Earth.

On Mars, you'll feel light on your feet, but still grounded. The gravity on the surface is a little more than a third that of Earth. Temperatures can reach a balmy 70 degrees Fahrenheit near the equator, though most of the time Martian temperature is eighty below zero. If you've already been to the Moon and back and you're ready for a new challenge—but aren't ready to devote your life to space travel as required by a trip to say, Pluto—then Mars is for you.

 **AT A GLANCE**

DIAMETER: A little over half of Earth's

MASS: 11 percent of Earth's

COLOR: Tan, brown, and rusty red

SPEED AROUND THE SUN: 54,000 miles per hour

GRAVITATIONAL PULL: A 150-pound person weighs
57 pounds

AIR QUALITY: Very thin, mostly carbon dioxide, with traces
of nitrogen, argon, oxygen, and carbon monoxide

MADE OUT OF: Rock

RINGS: None

MOONS: 2

TEMPERATURE (HIGH, LOW, AVERAGE): 95, -128,
-81 degrees Fahrenheit

DAY LENGTH: 24 hours and 40 minutes

YEAR LENGTH: 23.5 Earth months

AVERAGE DISTANCE FROM THE SUN: 142 million miles

DISTANCE FROM EARTH: 34 million to 249 million miles

TRAVEL TIME: About 200 Earth days for rendezvous

TEXT MESSAGE TO EARTH: 3 to 22 minutes

SEASONS: Frigid winters and chilly summers

WEATHER: Intermittent dust storms, occasional clouds

SUNSHINE: A little less than half as bright as on Earth

UNIQUE FEATURES: Solar system's largest volcano, Mount
Olympus, and largest canyon, Mariner Valley

GOOD FOR: Rock climbing and low-gravity hiking

Weather and Climate

If you want an escape from the sweltering summer heat, Mars is an ideal glacial getaway. About one and a half times as far from the sun than Earth, Mars is much colder.

The cycle of seasons on Mars is very similar to Earth's because the planets share an almost identical tilt, differing by less than two degrees. However, seasons are much milder on Mars because of its paltry atmosphere. There are no snowstorms, thunderstorms, or falling leaves (or trees, for that matter). The change of seasons is subtle. You may notice a difference in the way the sunlight hits the rock, the strength and direction of the winds, or the presence of clouds from one season to the next. The greatest transition happens at the poles, where the polar caps grow and wane with the change in sunlight.

Though the tilts are similar, Mars has a more elliptical orbit around the sun than Earth does. It's a common misconception that Earth's distance from the sun causes its seasons. The reason Earth's distance from the sun *doesn't* strongly affect its seasons is because its orbit is close to circular, so it's always about the same distance from the sun. On Mars, in addition to the tilt, the distance from the sun contributes to the quality of its seasons because of its more elongated orbit. Mars is farther from the sun when the southern hemisphere is in winter, so winter there is very cold. Its northern winter is a little milder because it's closer to the sun.

No matter the season, dust is a way of life on Mars. It's like a second skin that will coat your suit and cause trouble for the airtight seals of your rovers and habitats, and for the gears of machinery. Frequent dust storms can sometimes cover the whole planet, blocking out the sun. The best strategy if you're caught in one is to seek shelter in your habitat or vehicle and wait it out. Despite their foreboding appearance, dust storms of Mars look worse than they feel. The atmosphere is only about one one-hundredth as thick as Earth's, and so while they can make it hard to see or generate solar power, the winds feel more like a summer breeze than a gale. The exception is the very strong windstorms that sometimes occur during the changes in season, especially near the polar ice caps.

Occasionally you'll see clouds in the Martian sky. Made mostly out of water ice, they stand out in the orange skies because of their white color. Martian clouds are low, thin, and wispy. You may also be lucky enough to experience the fog of Mars. Like fog on Earth, it forms near the cool ground in low-lying areas, particularly in deep canyons such as Mariner Valley. Like its earthly counterpart, it fades away as the sun comes up.

Despite all the sand, you won't find anything like an Earth beach on Mars. Liquid water can't exist on the surface long because the pressure is too low—the same as what you'd find at an altitude over 19 miles high on Earth. Water evaporates very easily in these conditions, even at temperatures well below the freezing point of water on Earth. As such, flowing liquid water is a novelty, and tourists enjoy searching for these spontaneous and ephemeral flows. They're seasonal, appearing mostly during the summer months. You can find some of these natural flows at Hale crater, north of Argyre basin.

Outside of the plentiful accumulation of ice at the north and south poles, where it's preserved in permanently shaded craters, most water is below the surface. If you find any Martian water, don't drink it without purifying it first. It's salty, which can keep it from freezing, and often contaminated with chemicals called perchlorates. They're useful for making rocket fuel—and highly poisonous.

When to Go

Whenever you visit Mars, there will be sights to see. Visit in northern winter, when the polar ice cap is at peak size. Or go in time to witness either hemisphere's "endless summer." It's not truly endless, but about twice as long as Earth's summer—a pleasant bit of news for students and teachers who cherish summer breaks.

During northern summer, Mars is at its farthest point from the sun, and cloud watchers will see the most clouds near the equator. If you want to avoid cloudy skies entirely, visit during the southern hemisphere's spring and summer, but keep in mind that is also the peak season for dust storms.

Getting There

As with any vacation destination that is moving at a different speed than your point of origin, you'll need to carefully plan when to leave for Mars. The Earth is moving at 66,616 miles per hour, and Mars is moving at 53,843 miles per hour. Imagine your slower friend is running around a race track, and you're in an inner lane, about to toss her a ball. It doesn't make sense to throw it when she's on the other side of the track. You'll need to anticipate her motion and the travel time of the ball to figure out the best time to release it. In the same way, it is best to wait until Earth and Mars are in the best position before you depart.

You could leave *right now* for Mars—put its coordinates into your interplanetary positioning system and gas it, and then come to a screeching halt when you get there. We don't recommend that. A Hohmann transfer orbit, an ellipse whose farthest point from the sun is at Mars and closest point is at Earth, will use less energy than a spur-of-the-moment direct route, but you'll need to wait for a launch window, when the planets are in their best positions to start travel between them. This is when you and your friend's paths around the track are optimally lined up. For the simple Hohmann transfer orbit flight plan, these occur every twenty-five and a half months, or about once a Martian year. The farther in time you depart from your ideal launch window, the more fuel you'll need to get to Mars.

Once you've reached Mars, you'll have to wait at least eighteen months before the next opportunity to return home. And if you miss your return flight, you could be stuck on Mars for three and a half years while you wait for the planets to align again. This might not be the end of the world, as it means a longer vacation. Your boss won't be able to argue with you about why you need to extend your personal leave.

Another economical option for travel to Mars is to take a ride on an Aldrin cycler. It's like a subway line to Mars. Ships on an Aldrin cycler scuttle past Earth and Mars at regular intervals in a stable orbit, making use of gravity assists from both Earth and Mars to save fuel.

There is no limit to the number of ships that can share the orbit. The one-way trip from Earth to Mars on the cycler would take 147 days. A perk of riding an Aldrin cycler is that it is relatively inexpensive because the ship needs very little fuel to stay in orbit—just an occasional boost.

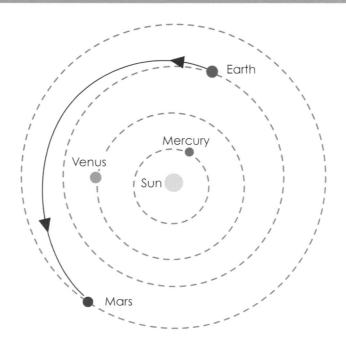

When You Arrive

From a distance, Mars looks like a tiny red spot in the sky. As you pull closer, the red will spread and you'll begin to see its features materialize. Look for the huge gash across the face of the planet, a canyon known as Mariner Valley. When you're ready to take a closer look, you can hop on a shuttle to the rocky surface. You'll be awed by Mars's red rocks. The color, not a fire-engine red but similar in tone to rust on Earth, is from iron oxide, hints that the planet once had flowing water. It looks a lot like Utah, but there is something different about the way the light strikes the rocks. You'll soon notice that in addition to rocks, Mars is covered with a fine dust, which gets on and in everything. The dust is kicked around by winds in an endless cycle, as there is little moisture to capture it and ferry it to the ground the way rain showers do on Earth. Instead, the dust stays airborne, imbuing the sky with its particular orangish color.

The rhythms of Mars are similar to Earth's. One day is only slightly longer on Mars, so if you've always wished there were just a little extra time in the day, you can finally get it. The Martian day is forty minutes longer than Earth's standard, giving you plenty of time to run that extra errand or linger just a bit longer at the dunes. In order to facilitate the transition from Earth time, upon arrival visitors trade in their Earth clocks for Martian ones that keep Mars time with hours, minutes, and seconds that are slightly longer (2.7 percent longer, to be precise) than their Earthly counterparts.

On Mars, the term *sol* refers to one Martian day. One sol is exactly twenty-four Martian hours long. The Martian analog of today is *tosol* and yesterday is *yestersol*. There are three options for tomorrow: *nextersol*, *morrowsol*, and *solmorrow*. Be prepared to be asked if you are on "soliday," the Martian equivalent of *holiday*. These terms were coined by NASA rover operators who had to find a way to mark the unusual passage of Martian time.

As soon as you land you'll probably want to call home to announce your safe arrival. Although in the scheme of the solar system you're close to home, it still takes time to relay messages back to Earth. Depending on the position of the planets, you'll wait six to forty-four minutes to receive a response to a text you send to Earth. Surfing the Internet is an exercise in patience.

Mount Olympus (foreground) is one of the most celebrated tourist destinations on Mars. Be sure to visit its trio of companion volcanoes, Ascraeus Mountain, Pavonis Mountain, and Arsia Mountain (middle, from left to right) and take a hike through the Labyrinth of the Night (top left).

ESA/DLR/FU BERLIN/JUSTIN COWART

Getting Around

You'll be off-roading in a rover for most of your travels on Mars. Choose a vehicle with robust wheels, since the sharp Martian rocks can easily lay waste to them. The most basic and economical buggies are similar to those used by the *Apollo* mission astronauts—small, completely open vehicles with seats that look like lawn chairs. You'll simply wear your space suit and strap yourself in. While great for a joy ride, these aren't good for an extended Martian road trip. If you want to cover some ground and really see the sights, you'll need a larger vehicle, a sealed RV that lets you carry your home with you. These vehicles can keep two passengers alive for up to fourteen days.

Air travel is possible on Mars, as long as the aircraft is specially designed to work in the ultrathin atmosphere. Gravity is weaker, which, apart from the fact that it also leads to a sparser atmosphere, helps planes stay airborne. Airships can carry only very light payloads, or must be very big to generate enough lift in the tenuous air.

Since low temperatures can freeze you in a Martian minute and there is no breathable air, a leisurely afternoon drive can turn into a perilous misadventure if your rover breaks down. In order to ensure your safety, it's advisable to follow the walk-back rule. Anytime you leave a protected habitat in a short-range vehicle, keep track of your power and oxygen levels and total travel time. Never venture so far that you can't walk to safety before your oxygen or power gives out. Conditions can change quickly.

Many rovers can operate autonomously, or by remote control. That said, there is nothing like the freedom of piloting a vehicle yourself. If you're planning to rent and drive your own vehicle during your vacation, you'll need specialized driver training. An excellent driver on Earth can discover that she's in hopelessly over her head when it comes to the demands of traversing between sharp rocks and dangerous cliffs, all without the help of smoothly paved roads.

Various models of rovers will differ from one another, but the general components often remain the same. A midrange rover will have a protective cover, a computer for navigation, a temperature regulation system, sensors to evaluate environmental conditions, robotic limbs for control, wheels or tracks for mobility, an energy source for power, and a communications system. No matter the model, none of them are very fast.

When choosing a vehicle, think about what you'll be carrying. Will you only require the bare bones of what you need to survive, or will you be toting along two tons of rare rock? If you're planning to carry heavy loads, you'll want to have lots of big tires to distribute the weight without sinking. You'll also need wheels that can deform under their weight instead of sticking solidly and inflexibly into the ground.

This brings us to the next point: Be sure to check the tires on any vehicle you buy or rent. Gas-filled rubber tires don't work on Mars, as the cold temperatures would cause them to shatter. Most likely your tires will be one of the pneumatic alternatives, which use spokes much

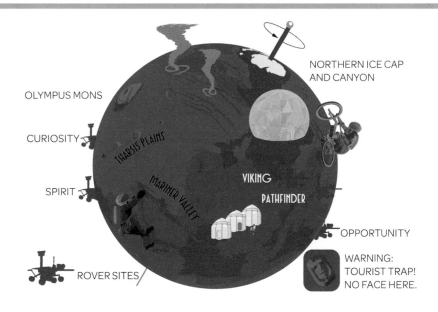

NORTHERN ICE CAP AND CANYON

OLYMPUS MONS

CURIOSITY

THARSIS PLAINS

SPIRIT

MARINER VALLEY

VIKING

PATHFINDER

OPPORTUNITY

WARNING: TOURIST TRAP! NO FACE HERE.

ROVER SITES

Mars offers adventures for every type of visitor, from hiker to history buff.

like those on bicycle wheels to provide the tires some give. Some vehicles use spiderlike hinged limbs with small wheels on the bottom. They can roll on the wheels like a traditional car, but can also walk over obstacles such as large boulders and deep sand. If you need to dig or drill, tools can be attached to one of the limber legs.

What to See

Mount Olympus (Olympus Mons)

The gentle giant of Mars, the long-extinct Mount Olympus is the tallest volcano in the solar system. From its base, its sixty-thousand-foot-high peak is 150 miles away, obscured by the nearby slopes. Summiting Mount Olympus is a highlight of any trip to Mars. The journey up the gradual slope is easy compared to the steeper mountains of Earth, but it's lengthy. Budget at least a month to complete the hike, and make sure you have a buffer of extra supplies. The cliff that surrounds Mount Olympus, marking its edge, is tumultuous. Approach the mountain from the south or east, where this initial barrier is less steep.

Syrtis Major

A dark stain on Mars, Syrtis Major was first plotted by Dutch mathematician Christiaan Huygens in 1659, and was later used by astronomers to track the planet's rotation. This led to its nickname, "the Hourglass Sea."

Today the souvenirs most popularly sold in this region are hourglasses filled with Syrtis Major's characteristic dark orange

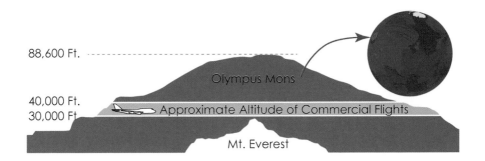

88,600 Ft.

Olympus Mons

40,000 Ft.
30,000 Ft.

Approximate Altitude of Commercial Flights

Mt. Everest

sands. Scientists once thought the feature was blue in color, and that it was overgrown with blue-green vegetation. However, this is just an optical illusion created by the contrast between the otherwise brighter and redder regions on the surface. In reality the dark color is due to the tint of its unweathered volcanic rock. The wind sweeps the brighter, freshly eroded sands away to other desert areas, leaving the darker rubble relatively bare. Nevertheless, Syrtis Major earned the additional nickname "Blue Scorpion" due to this misperception.

In the southwest section of the dark patch of Syrtis Major is Huygens crater, Mars's largest. It's about the size of New York State, and like many large craters, it features several concentric rings.

Hellas Plains (Hellas Planitia)

The oldest region of Mars is blemished with numerous craters left over from the early history of the solar system. Hellas shows up as a bright round area south of the dark patch of Syrtis Major. On cold mornings in the winter, it fills with a frosty haze, becoming bluish white in color. The effect is short-lived, burning off by midday.

Tharsis Plains/Tharsis Bulge (Tharsis Planitia)

The Tharsis plains make up a huge plateau larger than the area of the United States, formed from volcanic and tectonic processes. Sitting on top of it are three tall volcanoes, known collectively as the Tharsis Mountains (Tharsis Montes), and individually as Mount Ascraeus (Ascraeus Mons), Mount Arsia (Arsia Mons), and Mount Pavonis (Pavonis Mons). The peaks are higher than Mount Olympus, but only because the land surrounding the mountains on the Tharsis bulge is already at a high elevation. Mount Ascraeus is easier to climb than Olympus because the distance you'll need to cover is less and the slope isn't too steep. If you climb all the way to the top you'll be treated to the volcano's dormant center, a caldera with several areas where lava has cooled to pool-smooth rock.

Labyrinth of the Night (Noctis Labyrinthus)

This area is crosshatched with deep cracks flanking the western edge of Mariner Valley. The canyons form a maze, and you can lose yourself in them for days. As you wander, you may encounter landslides, dunes, and fascinating terraced rocks. The layered mesas will remind you of the rugged beauty of the badlands of South Dakota.

Arabia (Arabia Terra)

Polka-dotted with numerous blemishes, Arabia is ancient. Some of these blemishes are craters that were formed from impacts, and many others are evidence of volcanoes. Arabia has a vast field of

Stunning simulated view from the rim of the Black canyon of Mariner Valley

NASA/JPL/ASU/R. LUK

dark sand dunes that winds have sculpted into towering, seven-hundred-foot peaks. Some regions of Arabia show channels, which may be the dry remains of ancient riverbeds and streams.

Mariner Valley (Valles Marineris)

The climax of your Mars visit will surely be a trip to Mariner Valley, near the Martian equator. Named after the *Mariner 9* Mars orbiter that discovered the canyon in the early 1970s, it's the grandest canyon in the solar system. This massive rift valley is twenty-five miles across and seven miles deep—more than four times deeper than the Grand Canyon. It stretches about a quarter of the way around Mars—as long as the United States is wide.

There are several picturesque views at various points along the rim of the canyon. One of the most popular is at the intersection of Ophir canyon (Ophir Chasma) and Candor canyon (Candor Chasma), where you will find yourself nearly surrounded by steep cliffs. Another vista is in the middle of the Black canyon (Melas Chasma), or from the deepest point in Coprates canyon (Coprates Chasma), where the walls are thirty miles away and the summit towers almost a mile above you.

One thing people find astonishing about the views of Mariner Valley is that unlike in canyons on Earth, when the dust has subsided there is less haze in the air, because the atmosphere is thinner. Like a newly cleaned window you didn't realize was dirty, the landscape can look surprisingly crisp and clear.

Polar Ice Caps

Those who are concerned about access to water like to travel to the poles, but these areas can be cold, especially in the winter. The swirling spiral shapes of the polar ice caps were formed by winds powered by Mars's rotation and gravity. The northern cap is much bigger than the southern cap, and is made mostly of water ice with a layer of carbon dioxide ice on top. It houses the Northern canyon (Chasma Boreale), whose sculpted walls are layered like ribbon candy. In winter, carbon dioxide from the atmosphere freezes to the top of water ice, building up the massive ice sheet. In the summer, when the sun hits it, the carbon dioxide ice layer warms and vaporizes, leaving the water ice exposed and ready for harvest. There are strong winds at the poles during the change of seasons.

Historical Sites

Visit the final resting places of the robotic explorers of Mars: NASA's Curiosity, Spirit, and Opportunity rovers. Curiosity is located in Gale crater with a lovely vista of Mount Sharp, while Spirit ended its mission in an area called Troy, on the western side of a region called Home Plate. Opportunity's resting place is near Endeavour crater, in Meridiani plains. While the tracks have been blown over by dust, it's fun to retrace the routes that the rovers took. In the case of Opportunity, the path is longer than a marathon.

A stunning vista of Mount Sharp, along the historic path of the Curiosity rover.

NASA/JPL-CALTECH/MSSS/J. GRCEVICH

Activities

Enjoy the Martian Skies

On Mars, the landscape is reminiscent of the arid deserts of Earth, but the familiarity will fade as soon as you look up and see the rusty sky. On Earth, the sky is blue because of the way light scatters off air. The effect is different on Mars, where the hue of the sky comes from light scattering off dust particles rather than the sparse air. The area around the sun in the sky is much brighter and bluer than the rest of the sky.

The gorgeous red tone of Mars that you see in all those idyllic promotional images is a bit of a stretch. The reds aren't quite as pure in person as they are on the postcards. Upon closer inspection of the rocks, you'll find they are a delicious shade of golden, tan, and brown.

Sunsets on Mars are delightfully alien. Since the planet is farther from the sun, the sun appears smaller from Mars than it does from Earth. The colors during sunset are the reverse of the typical colors of Earth's sky: The sky far from the sun is reddish, while the area around the sun is blue. The light scatters off the dust and creates a blue hot air balloon shape with the sun as its basket. Because Mars rotates around its axis at about the same rate as Earth, the sun sets at about the same rate as well, but the dusty sky reflects light from the already-set sun, making twilight last much longer. In a dust storm you won't be able to see the sun as it sets. Instead, the sun will sink into the murky haze.

While the daytime sky might seem exotic, the nighttime Martian sky will be familiar: black and dotted with stars. You'll notice all the usual constellations you would see from Earth, except for the presence of a new star, which is not a star at all. Earth, third planet from the sun, appears as a blue-tinged point of light from the surface of Mars. You may see it either in the morning or in the evening Martian sky. Even Earth's moon is also visible as a point of light—a dimmer companion star to the bright Earth.

Though the constellations are the same, Martian sky watchers may notice that the stars move differently than they do on Earth. Back home, the North Star, Polaris, is lined up with Earth's rotational axis, perched directly above the top of the planet's North pole. It stays in the same place in the sky all night, while the rest of the stars move slowly about it. Because Mars's axis is tilted in a different direction than Earth's, it points at a different north star. Mars's north star is dim and hard to see, in an area between the constellations Cygnus (the swan) and Cepheus (the king). There is a bright south star for Mars, though—Kappa Velorum, in the constellation Vela (the sail). Visitors to Mars's southern hemisphere will see the sky appear to rotate around this star over the course of the night.

Skydive

Skydiving the Martian skies is a much riskier sport than it is back home. On Earth, you eventually reach a constant speed because you are slowed down by the air. This speed is called terminal velocity, and it's 124 miles per hour. Back home, you'll never free-fall faster than that.

On Mars, because the atmosphere is so much less dense, your final velocity will be five times faster than that. You'll have to use a series of parachutes, pull them earlier, and they must be huge to slow you down enough. There is no thrill like it on Earth.

Rock Climb

The dramatic cliffs of Mariner Valley are the perfect place to try out your rock-climbing skills. If you're not keen to climb upward, try rappelling down the walls of the canyon at its deepest point, in Black canyon.

It can take a while to become accustomed to climbing in a bulky space suit at low gravity. Some newbies believe that because of the low gravity you can't get hurt if you fall, but that's a myth. You may fall slower at first, but by the time you hit the floor of Mariner Valley, you'll end up like a bug on a windshield.

Chase Dust Devils

Even when there aren't dust storms, the Martian winds can whip up action. Swirling cones of dust travel over the surface like tornadoes. Compared to their earthly counterparts, these Martian dust devils can quickly spiral into gargantuan towers, stretching a mile high and more than five hundred feet wide. If you enter one, the wind won't be too strong, but the dust particles will be moving very fast and may scour and scratch the face shield of your space suit. Inside the dust devil, you'll see tiny bolts of lightning striking you from charged dust.

Bike

Mars is one of the few places in the solar system where you can bike. Since you'll be biking off-road, Martian bike tires are thick so that they can ride above the rugged surface instead of sinking into the dust. They are also textured, so you can get a good grip on the rocks. Because of the cold, tires are not inflated tubes of rubber, which would become brittle and shatter easily, but instead have springy spokes.

Steering is more difficult in the low gravity of Mars. In order to turn, you'll have to go slower, and turn less sharply. It's also tricky to accelerate quickly without popping a wheelie, since less gravity means the wheels have less friction with the ground. The good news is that on a paved, well-banked road, bikers racing on Mars can go *much* faster because there is barely any air resistance.

Juggle

If you're learning to juggle, Mars is a great place to do it. About one-third Earth's gravity means you'll throw the balls higher with the same force, making for a slow but dramatic show. Even if you throw them to a typical Earth juggling height, they will take longer to fall. This is great news if you don't have the best reflexes—you'll have plenty of time to master the motions.

What's Nearby

Mars has two small moons, Phobos and Deimos, perfect for side trips. Phobos zips around Mars in just nine hours, three times per Martian day. Deimos circles the planet every thirty hours, but orbits in the opposite direction. Depending on where you are on Mars's surface, the moons will pass in front of the sun several times a year. These so-called transits can be thought of as mini-eclipses, but only a tiny part of the sun is blocked. On rare occasions you can catch both moons passing in front of the sun at the same time—a double eclipse.

Phobos

Though it's named after the Greek god of fear, this moon is a delight to visit. It's just forty-three miles around, so you can explore all of its wonders in only a few days. If you're an ultrarunner experienced in low-gravity running, you could run around the whole moon in a day. More fun than running is jumping—you could just clear Earth's tallest building, the 2,722-foot-high Burj Khalifa in Dubai, in a single epic leap.

Visit Mars's larger moon, where you could jump over Earth's tallest building.

HIRISE/MRO/LPL (U. ARIZONA)/NASA

Phobos orbits closer to Mars than any other moon to its planet. Though Phobos looks small from Mars—about a quarter of the size of a full moon on Earth—from the surface of Phobos, Mars looms large, at eighty-five times the size of the full moon from Earth. Phobos is predicted to fall apart sometime in the next 30 to 50 million years due to stretching and squishing from Mars's gravity. These tidal forces from Mars cause shallow grooves to form over the surface of Phobos, sometimes referred to as the moon's stretch marks. The inside of this satellite is only loosely held together by gravity and is encased in a dusty outer shell.

Deimos

The twin of Phobos, Deimos, whose name means "panic," is the smaller of Mars's moons, at just twenty-four miles around. Its tiny mass means gravity on the surface is very low, thousands of times lower than on Earth. With an escape velocity of only 12.5 miles per hour, just about anybody can throw a baseball hard enough to put it into orbit. Don't hit a landing ship. Walking is difficult when the slightest jump sends you flying. You'll have a lovely view of Mars, which will be thirty-three times as large as the full moon appears from Earth. From Mars, Deimos looks slightly brighter than Venus does from Earth.

Asteroid Belt

The early solar system was a messy place, with lots of objects, big and small, colliding and coalescing. While today we may think of the solar system as a neat collection of eight planets in their orderly orbits around the sun, there are still cosmic remnants scattered in between. One type of leftover debris are the asteroids—space rocks. Their low gravity and irregular shapes make them fascinating places for a quick stopover. While asteroids can be found anywhere in the solar system, many are located between the orbits of Mars and Jupiter, in a region known as the asteroid belt.

The asteroid belt is one of those places that have a dangerous reputation but don't deserve it. The distance between asteroids is huge, such that you usually only see the one you're visiting, and if you're just passing through, you can navigate the belt with your eyes closed. There isn't that much stuff there—if you collected all the asteroids in the belt, you'd end up with something less than one-fourth the mass of Pluto.

Ceres

A single object, Ceres, comprises one-third of the mass of the asteroid belt. It's the only dwarf planet in the asteroid belt, and no trip to the belt is complete without a stop to admire its brilliant bright white spots. The brightest of these spots is at the center of the large Occator crater. The spots are made of Epsom salts from briny ice that evaporated. Collect some to give to your friends back home; it makes for a relaxing soak in the bath. Ceres is named after the Roman goddess of harvests and corn, which is also where the word *cereal* comes from. Next time you're eating corn flakes, think of your next trip to this dwarf planet.

Vesta

At one-fourteenth the mass of Pluto, Vesta is the biggest object in the asteroid belt which is not also a dwarf planet. When visiting Vesta, look for the grooves that encircle it at the equator, like the stretch marks of Phobos, but here thought to be caused by a giant impact. If you're feeling adventurous, climb the central mountain of Rheasilvia crater, which provides further evidence of Vesta's tumultuous past. At 13.7 miles high, it is the tallest mountain in the solar system.

JUPITER

TOUR THE NORTHERN LIGHTS

Destination ♃ Jupiter

Jupiter is the undisputed king of the planets. The first gas giant you'll visit in the outer solar system, he is there to greet you with a seeming stillness. The tranquility of his pearly banded silhouette is an illusion. Jupiter is a planet of storms, a force of nature, with a mass far greater than that of the rest of the planets in the solar system combined. His power is intoxicating.

If you are the sort of person who is drawn to unbridled chaos, Jupiter will not disappoint you. Here is a planet so large, a single one of its storms would swallow Earth. At the top of its clouds, its gravity is two and a half times greater than ours, and the magnetic field is twenty thousand times stronger. Jupiter's magnetosphere extends almost to the orbit of Saturn, bathing its moons in radiation along the way.

Though it's easy to become hypnotized by the planet's sand sculpture cloudbursts, it's the moons that will seduce you. The Jovian system is a solar system within a solar system, hosting satellites as varied as—and some even bigger than—planets.

♃ AT A GLANCE

DIAMETER: More than 11 times Earth's

MASS: 318 times Earth's

COLOR: Swirling reds, browns, burnt orange, rust

SPEED AROUND THE SUN: About 29,000 miles per hour

GRAVITATIONAL PULL: A 150-pound person weighs 355 pounds

AIR QUALITY: Thick, 90 percent hydrogen, 10 percent helium, with some methane and ammonia

MADE OUT OF: Gas

RINGS: Yes

MOONS: 67

TEMPERATURE AT 1 BAR: -162 degrees Fahrenheit

DAY LENGTH: 9 hours and 54 minutes

YEAR LENGTH: About 12 Earth years

AVERAGE DISTANCE FROM THE SUN: 484 million miles

DISTANCE FROM EARTH: 366 million to 602 million miles

TRAVEL TIME: 1.5 Earth years for flyby

TEXT MESSAGE TO EARTH: 33 to 54 minutes

SEASONS: None

WEATHER: Intense

SUNSHINE: Less than 4 percent of Earth's

UNIQUE FEATURES: Great Red Spot, active moons

GOOD FOR: Bodybuilding, watching auroras, moon hopping

Weather and Climate

Pack your best storm gear, and prepare to feel the wind on your face(plate). Weather on Jupiter is never dull. Everything on this gas giant is more extreme than on Earth. Wind speeds are 120 miles per hour higher than record gusts back home. Storms here can last for decades, and the most famous, the hurricane known as the Great Red Spot, has been raging for hundreds of years. The skies are full of lightning a thousand times more powerful than on Earth, and thunder races across the sky four times faster. If you're used to measuring the distance from a lightning strike by counting seconds from flash to boom, relax. The storm will be four times farther away than you think on Jupiter. The thunder is unrecognizable—not a low rumble, but an eerie screech, shifted in pitch by the abundant hydrogen and helium in the atmosphere.

Jupiter is the fastest-rotating planet. At the top of the clouds near the equator, the length of the day is only ten hours. Don't feel bad if you sleep away an entire Jovian day. It's perfectly normal. Because the planet has no solid ground, the length of day varies between the equator and the poles. The gases move slower at the poles, and you'll gain a few minutes per day when you're near them. The concept of a day is flexible in this amorphous, spinning sphere of turbulence. But that's half the fun.

It's cold out here, 500 million miles from the sun, where it is only 4 percent as bright. Though Jupiter is virtually made of storms, conditions remain steady throughout the twelve-Earth-year-long journey around the sun, because Jupiter has a very small tilt of only a few degrees. The temperatures vary as you navigate up or down through Jupiter's bright clouds. The usual forecast calls for high winds and temperatures in the -160s at one bar level, the region where the atmospheric pressure is similar to Earth's at sea level. Not warm by any stretch, but there are even colder places farther out.

When to Go

You might think it's best to launch when Jupiter and Earth are at their closest, but if you did that, by the time you reached Jupiter's orbit the planet would be long gone. Earth laps Jupiter easily because of its smaller orbit and faster speed, so you'll have one launch window each year to catch a Hohmann transfer orbit.

There are no seasons, so the weather is the same on Jupiter any time of year. The Great Red Spot is shrinking, and could be gone within decades, though no one knows for sure. Best to not press your luck—book a trip as soon as you can.

Most visitors prefer to limit their time on Jupiter and its inner moons, where the radiation levels are formidable. Because of that, you'll probably spend more time getting there than vacationing there.

Getting There

If you're just passing by Jupiter for a gravity assist on your way to another planet, you can make it there in a few years with a regular old rocket. If you want to stay and see the sights, you'll have to take your time slowing down to match Jupiter's average speed of 29,214 miles per hour. You'll arrive five or six years after you leave Earth.

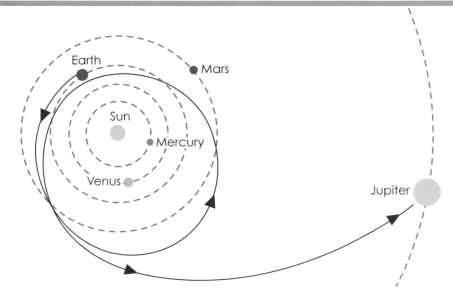

Jupiter is often used as a gravity assist to slingshot you to other vacation destinations.

If you're not in a hurry and want to minimize waste, you might consider riding an ion propulsion rocket. These rockets give a small push over long periods of time, causing very gradual changes in your spacecraft's speed. While your car can easily go from 0 to 60 miles per hour, an ion engine could take four days to do the same.

When You Arrive

Before you see Jupiter, your radio will hear it. A few weeks from arrival, about 3.5 million miles from landing, you will encounter the boundary of Jupiter's magnetosphere. As you pass it, your radio will capture the sound of supersonic solar winds crashing into the magnetosphere, which heats and slows high-energy particles traveling from the sun. The sound is eerie, screechy, and haunting, reminding you that you are hundreds of millions of miles from home.

The spooky feeling is temporary. As you approach the god of the planets, you'll feel a growing sense of comfort as you see the distant orange-brown orb slowly getting bigger until it seems impossibly huge. Nothing will prepare you for the experience of seeing it firsthand. Some say it reminds them of an abstract watercolor; others, a hot fudge sundae. Settle in and take a moment to enjoy the psychedelic panorama. It's easy to lose time as you watch the striated gases of Jupiter slow dance.

Getting Around

In orbit, you'll experience the microgravity you're accustomed to from your flight. Studying Jupiter's atmosphere from a high orbit is a perfectly lovely way to spend your vacation. If you're brave enough to take a closer look at its frenzied storms, a rocket ship can ferry you to high-altitude airships where you'll start to feel your weight. This new heaviness may feel oppressive. Even if your airship can stay near the highest cloud tops, you'll feel several times Earth's gravity bearing down on you.

If you descend to where the pressure of Jupiter's atmosphere is the same as Earth's, you'll weigh two and a half times what you do back home. The trick is to avoid moving around too much at first. Lie back and enjoy the weight. Some say it feels like a firm hug, evoking feelings of safety and calm. If you can avoid passing out long enough to master movement in high gravity, you'll quickly build muscle, which will come in handy for the ride home.

Before you get too relaxed cruising through Jupiter's skies, make sure you're sufficiently protected from the radiation. Water or lead shields can protect you and your equipment. Titanium is a lighter choice, but less effective. Savvy travelers carry a dosimeter, which measures the amount of harmful ionizing radiation you've been exposed to.

Your main mode of transportation will be an airship. It's harder to stay afloat in Jupiter's superlight atmosphere than back on Earth. You'll need to use hot pure hydrogen as your lifting gas. A helium balloon sinks in Jupiter's hydrogen-rich atmosphere. Be careful not to mix hydrogen with oxygen or the results can be explosive.

On your vacation you'll spend plenty of time sightseeing on Jupiter's moons. Before you depart for them, make sure you're really ready to leave Jupiter behind. It is very difficult to leave, both emotionally and physically. It's a massive planet, and launching from it takes five and a half times more energy than launching from Earth. This is made more challenging by the fact that you have to launch from the atmosphere.

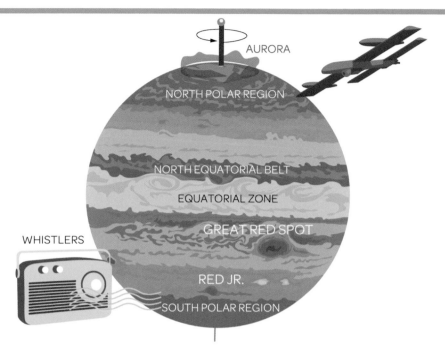

AURORA

NORTH POLAR REGION

NORTH EQUATORIAL BELT

EQUATORIAL ZONE

GREAT RED SPOT

WHISTLERS

RED JR.

SOUTH POLAR REGION

You'll be enraptured by a close-up view of Jupiter's gaseous displays.

What to See

The Great Red Spot

The first thing you'll want to see is the Great Red Spot, which attracts flocks of sightseers. This enormous storm is twelve thousand miles wide, and could easily fit our entire Earth within it. You shouldn't have trouble finding it, since it's always at the same latitude, south of the equator in a region known as the South Equatorial Belt. We don't know why it's there, or how old it is—just that it's been there ever since humans have been looking at it. On Earth, it would be a Category 20 hurricane. Look for bright white ovals that sometimes appear nearby. Occasionally they get sucked into the colossal vortex.

The Great Belts and Zones

Once you've gotten your fill of Jupiter's churning red eye, broaden your view and take in the dizzying dynamics of the planet's windy cloud bands. While Jupiter as a whole spins, the winds are also doing

Only the brave dare to fly through the swirling larger-than-Earth hurricane known as the Great Red Spot.

NASA/JPL/BJÖRN JÓNSSON

their own thing in addition to the global motion. Each stripe represents a powerful jet stream with winds greater than 200 miles per hour. That would knock you off your feet, if you had anywhere to stand. Jupiter has two types of bands: belts and zones. The clouds in the zones are light in color, with shimmering high-altitude ammonia ice clouds. Winds in the zones blow in the same direction Jupiter rotates. The dark belts, on the other hand, have lower-altitude clouds and winds that usually blow opposite Jupiter's rotation. Be careful at the boundaries between zones and belts, where winds change direction and your ride can be bumpy. Look closely, and over time you'll see the ribbons of gas change color. Areas that are normally bright white can give way to yellowish orange, dark brownish yellow, or red as gases mix and winds collide.

Visit the Grand Auroras

You'll pass multiple belts and zones of gas as you fly toward the north pole, where you'll witness the best aurora you've ever seen. On Earth, the appearance of northern or southern lights is sporadic, and you have to be in the right place at the right time to catch the rippling luminescence. On Jupiter, the auroras are always on, are always at the poles, and are one thousand times more powerful than on Earth. Though the radiation near Jupiter is dangerous, it is responsible for the solar system's most impressive auroras, which arise when charged particles crash into the upper atmosphere. Intense purple jumps through the clouds, lighting up the entire sky in a supernatural glow.

Look for more auroras on Jupiter's moon Ganymede, which has a small magnetic field of its own. The red, green, and purple curtains of light span more than twelve hundred miles. If you could stand

Jupiter's impressive auroras, pictured here in false color, will awe even the most jaded space traveler.

NASA/ESA/HUBBLE

under them, they would fill the entire sky, moving at 10,000 miles per hour. They practically roar.

The North Pole

If you're a fan of Vincent van Gogh, you'll love Jupiter's north polar region. Here the usual bands and zones of Jupiter dissipate, leaving a sky of chaos. Impressionistic swirls form from mixing atmospheric layers and fluffy clouds, and you'll notice a bluish tint compared with the usual hues that you'll encounter while touring Jupiter. Don't expect a smooth ride—the atmosphere here is dotted with violent storms, like a churning minefield.

Activities

Tour the Ghostly Rings

Fans of Jupiter's rings can get a little defensive if you compare them with the extravagant rings of Saturn, but those who appreciate subtle beauty will enjoy a Jupiter ring tour. Dark, shadowy, and thin with a hint of red, they are made up of tiny dust particles. The main ring is

Jupiter's north pole has been compared to Vincent van Gogh's The Starry Night.

NASA/JPL-CALTECH/SWRI/MSSS

mostly transparent, and the gossamer rings that orbit farther out are even more so. Even if you fly right through them, you'll barely notice anything.

Listen to Trippy Space Sounds

You can tune your radio to the sounds of Jupiter's magnetosphere. They come and go every hour, like aural poltergeists. Though invisible to the human eye, the magnetosphere is one of the greatest wonders of the solar system, stretching for 450 million miles. If it were visible in Earth's night sky, it would take up the space of five full moons.

It arises from the interaction between Jupiter's magnetic field—generated from a large layer of metallic hydrogen coursing in its interior—and the solar wind, which delivers a constant flow of electrically charged particles to the neighborhood. The solar wind crashes into the magnetic field, creating high concentrations of radiation. Jupiter's radiation belts are not as benign as Earth's Van Allen belts because its magnetic field is ten to twenty times stronger than Earth's. Though Jupiter is far from the sun, its magnetic field is very good at trapping charged particles from it, meaning the planet has some of the highest radiation levels in the solar system. The broadcast may remind you of lion roars, whistles, hisses, woodpeckers, or waves crashing on the beach. Rest assured, there are no lions on Jupiter—just an electromagnetic jungle in the magnetosphere.

Dive the Atmosphere

Plunge deep into the depths of Jupiter's exotic atmosphere, where you'll encounter materials behaving in strange ways. In the strong gravity and thin air of Jupiter's upper layers, you'll drop faster than you would on Earth. Ammonia clouds hover where the pressure is similar to Earth's. A bit deeper you'll get a whiff of ammonium hydrosulfide clouds. Below that is a layer of white clouds, made of the familiar water vapor you're used to. The water clouds clash with the ammonia layer, sparking lightning.

By the time you've descended to where the pressure is ten times what it is on Earth's surface—equivalent to a three-hundred-foot dive into the ocean—the temperature reaches 150 degrees. Here it's already pitch-black from the thick layers of cloud above you. As you descend, temperatures continue to creep up. Fall for a few hours into the abyss and it gets hot enough to melt aluminum. Soon after that you'll reach pressures a thousand times that on the surface of Earth. It's like arriving at the bottom of the Mariana Trench, seven miles below sea level. Your vessel may survive to this point, but if you continue on, there is no hope for you. The pressures will only increase as you fall, and your ship will soon implode. Ten hours after you began your deep dive, even the titanium of your craft will have melted and then evaporated, becoming one with Father Jupiter. We recommend you turn back before that happens.

But what if you had an impossibly strong vessel and could continue? The upper atmosphere is much less dense than Earth's because it's made up of light hydrogen. As you sink, the soup of gas gets denser. When the pressure is half a million times that on Earth, you'll notice that you're surrounded by something more like liquid than gas: liquid hydrogen. Keep going and you'll reach metallic hydrogen, which is squeezed so tightly the electrons have been popped free of their atoms and roam loose. You can move through the metal seas as easily as through water, but visibility is very low. There's a rumor that around here, it rains diamonds.

Rent a Probe

If you're nervous about the iffy weather of Jupiter, you might consider crossing the chaotic expanse of its violent atmosphere with your own probe. From a control center on a nearby moon, you can rent one loaded with cameras and sensors. You'll see the planet's bizarre colors and storm systems close-up through a semi-live feed.

If you get good at navigating, you'll be able to make your probe last as long as possible. You can steer it and do your best to dodge lightning and radiation flares. Just remember that the probe will eventually implode if you go too deep. If this happens, you're unlikely to get your rental deposit back.

Some people get quite addicted to these robotic explorers, hiring probe after probe until they've traversed the whole atmosphere, only to discover new spots and storms. The exploration on Jupiter is never ending.

Watch Comets Crash

The comet Shoemaker-Levy 9 ventured too close to Jupiter in 1994, crashing into it at 134,000 miles per hour. The energy of the impact was six hundred times as powerful as all the world's nuclear weapons exploding at once. The largest pieces were more than a mile in diameter, and opened huge rifts in the clouds, revealing the layers below for months. Jupiter's whopping gravitational pull ensures many more asteroids hit it than hit Earth, and with greater force. Impacts can create huge plumes above the cloud layer and fast-moving ripples in the atmosphere that look like the waves from a pebble dropped in a pool of water.

What's Nearby

Jupiter's sixty-seven moons offer a sampling of the solar system's most extraordinary landscapes, from the volcanoes and subterranean oceans of the larger moons to the odd shapes and topsy-turvy orbital angles of the smaller moons. Some are barely a mile wide, and perfect for quick low-gravity hikes and moon hopping.

Jupiter's moon Io is covered in volcanoes.

NASA/JPL/SPACE SCIENCE INSTITUTE

You'll be dying to swing by the famous Galilean moons, which lie just beyond Jupiter's four closest moons. The set of satellites, named after characters from Greek mythology—Io, Europa, Ganymede, and Callisto—observed by Galileo in 1610, were the first moons discovered to orbit another planet. All but the farthest out, Callisto, have hazardous radiation environments. Make sure to keep your personal radiation detector close by, and double-check your radiation shielding before venturing outside protected ships and habitats. Only on sweet Callisto, a quiet, rocky moon a million miles from Jupiter (four times farther than the Moon is from Earth), can you escape "radiation anxiety," a condition that plagues many visitors who worry about the sea of radiation washing over them.

The moons move in delightful synchrony, orbiting their mother planet in a pleasing pattern. All of the Galilean moons are tidally locked, showing only one face to Jupiter as they circle around. The closest in, Io, goes around Jupiter in just forty-two hours. Next is

Moon-hop Jupiter's four largest and most fascinating satellites, known as the Galilean moons.

Europa, dancing around in twice that time, or three and a half Earth days. Ganymede takes four times as long as Europa, and Callisto takes the longest, almost seventeen Earth days.

Anytime is a good time to take off for these moons, since they are all locked in permanent equinox. Getting to know these fascinating spheres is sure to be a highlight of your trip to Jupiter.

Io

Orbiting Io from a safe distance above, you'll witness a beastly mess of orange, brown, yellow, and black. You might even think it resembles an overdone pizza. The moon is a volcano lover's Shangri-la, saturated with volcanoes, geysers, and lava fields, and home to the most magnificent eruptions in the solar system.

Radiation is higher on Io than on any other moon because it's smack-dab in the middle of one of Jupiter's radiation belts. Intense blasts from its volcanoes heave sodium and sulfur ions high into space, leaving a trail of ionized molecules that combine with Jupiter's powerful magnetic field to generate a ring of charged plasma around the planet. This plasma ring seeds the dazzling auroras at the poles. Jupiter's magnetic field also induces an electric field across Io, generating 3 million amps of current, which in turn sparks lightning in Jupiter's atmosphere.

The volcanoes of Io are fashioned in fissures of every shape and size. You'll see fiery lakes, lava rivers, and calderas—volcanic hills with a caved-in center. Huge volcanoes unload umbrella-shaped plumes of dust and gas into space, decorating the landscape with white, yellow, and red. After an explosion, sulfur dioxide frosts form, strangely reminding you of freshly fallen snow on a cold winter night.

One hundred calderas are erupting on Io at any given moment. It's the powerful tidal forces generated by the moon's motions with respect to Jupiter and its sister moons Europa and Ganymede that have shaped this fiery landscape. Three-hundred-foot tides are not unusual. The surface cracks as it seeks to release pent-up energy, belching forth a continuous stream of sulfur dioxide gas and lava.

Walking around firey fields is a pretty dodgy activity, though the lava is less of a threat than the radiation on Io. Unshielded visitors are exposed to a dose high enough to swiftly kill them. You'll want to keep a meticulous tally of your radiation exposure. Pack plenty of water, because all the volcanic activity has turned the moon arid. Prepare to navigate unstable ground prone to cracking. You'll hop from rock to rock over hot ash, sidestepping hidden geyser holes that could easily cut your vacation short.

You'll start your tour of Io at Loki, a gigantic lava lake named for the shape-shifting Norse god. From above it looks like a horseshoe, with dark liquid hugging a raised area of land. At 126 miles in diameter, Loki is Io's largest volcanic depression. Because of its size, it's more apt to call it a lava sea rather than a lava lake. Whatever you call it, just don't get too close to its edge, because glowing rock on its banks breaks off and falls into the fiery pit. Heat from Loki is viewable from Earth. Let your friends back home know exactly when you'll be there so they can look at it through a telescope and know you're having the adventure of a lifetime. The body of lava from Loki puts out more heat than all of Earth's volcanoes combined.

Southwest of Loki is Ra, a gorgeous volcano more than half a mile high. Veiny canyons sprawl from its central pool. It's known for its low-viscosity lava flows and high eruption rates.

Fourteen hundred miles east of Loki, about the distance

from San Francisco to Denver, is Pele, named for the fire goddess who created the Hawaiian Islands. It is surrounded by a red ring roughly the size of Alaska. In its center is a nineteen-mile-wide lake of lava. This is the site where human observers first realized Io was geologically active in 1979, when the *Voyager I* spacecraft, a ship passing in the night, captured an image of a 190-mile-tall volcanic plume.

Volcano tours are dangerous. You may encounter hot ejecta, avalanches, and toxic gases. While you may have the opportunity to see fire fountains, which are typically safe to view from a distance, note that an eruption of any size can happen at any time without warning.

From Pele you can head south to a huge mesa called the Danube plateau, named after the Danube River, which Zeus's lover Io passed through. The three-mile-tall mountain is 158 miles wide, and split down the middle by a canyon almost fifteen miles wide in places. You can hike above the canyon for striking views, but proceed with caution; the western rim is susceptible to landslides.

About two thousand miles due east you'll encounter Prometheus. Named for the Greek god who gave mortals fire, Prometheus looks like it's been erupting for decades. It's a relatively slow but consistent volcano surrounded by flowing lava fields and a bright plume of sulfur dioxide residue. It throws lava up 250 miles high. If you were to orbit the moon at that height, you might get an unexpected up-close glimpse of it outside your window.

Europa

A frigid ice moon slightly bigger than Pluto, the sixth moon from Jupiter is a haven for ice climbers, deep-sea divers, and anyone inter-

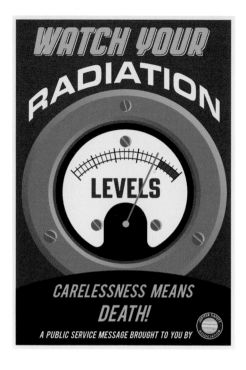

ested in extraterrestrial biology. Europa is geologically active, with geysers, ice volcanoes, and one-hundred-foot tides.

Looking down on Europa, you'll see a white surface cut with strange marks. These are huge cracks in the many miles–deep ice sheet covering the entire surface. The low gravity, 13 percent of Earth's, sends erupting ice plumes high into the sky. Below the ice is a sea with more water than all the oceans of Earth.

You'll want to watch your personal radiation levels as you travel Europa. Treat excursions to the surface carefully, the way divers time and track their depth when exploring Earth's oceans to avoid decompression sickness. At the surface you risk exposure that could kill you after just a few days. It's safer to stay under the ice, which is a natural shelter against radiation.

The moon is awash with water ice, which you can process for drinking or fuel. For your first drink, we recommend partaking in a traditional tea ceremony, a Japanese-style tea ritual with NASA-accented serving ware first imagined by Earth-based artist Tom Sachs, which will help keep you warm in the frigid hundreds-below-zero temperatures. At these temperatures, the ice that covers Europa behaves like rock.

Your first stop on the ice moon will be Pwyll crater, named for the hero of a medieval Welsh tale. It's a prominent depression carved into the ice on the Jupiter-facing side. The sixteen-mile-wide crater has a two-thousand-foot-tall central peak, which towers over its degraded one-thousand-foot rim. Many of Europa's craters, all named for characters of Celtic myths, are withered like this in some way.

From Pwyll crater you might consider drilling through the miles-deep ice to explore the hydrothermal vents of Europa's subsurface ocean. You could hire a private submarine, take diving lessons, or navigate the waters with hydrobots. Alternatively, continue directly north to Conamara Chaos, an ice form named after a region in western Ireland. It's one of five *chaos* areas on the surface of Europa—a mishmash of ice ridges, cracks, and plains. The relatively young feature was formed when the crust of the moon shifted and water from below welled up over the surface, breaking up blocks of ice, melting, and refreezing them.

EUROPA'S DEEP SEA EXCURSIONS WITH GUERILLA SCIENCE

Dive the Icy Ocean

Southeast of Pwyll crater is Agenor Linea, the San Andreas fault of Europa. It's a bright, 879-mile-long band twelve miles across. The smooth track is an ideal setting for snow rover racing.

Reach the end of Agenor Linea and you'll arrive at Thrace Macula, Europa's largest dark spot. Take a moment to remember the giant black monolith featured in science fiction writer Arthur C. Clarke's *2001: A Space Odyssey*, a book that cast Europa as a sacred satellite that may harbor intelligent life.

To the west of Thrace Macula is Thera Macula. There's a subglacial lake there, the perfect place to relax for a few days before your next adventure. Take a hot sauna and plunge into freezing baths fed by the lake. Before you finish your tour, don't forget to collect some Europan sea salt. Try it on a dish of chocolate ice cream for a true Europan culinary experience.

Ganymede

With an orbit 251,000 miles from Europa, Ganymede is a strange ice moon with elegant patterns of light swirls and fragmented dark terrain. It circles Jupiter half as fast as Europa, so you can hop over as Europa laps it. It's a dreamy moon, a retreat for romantics, with glittering craters dotting its patchy flowing surface like hundreds of tiny stars. A sprawling thin frost cap covers its north pole. It's the biggest moon in the solar system. If you spend at least a week here, you'll enjoy a 360-degree view of Jupiter as it completes its orbit.

When visiting, you'll soon become very familiar with strange grooved features called sulci, reminiscent of the brain's surface, crisscrossing the landscape. They rise two thousand feet into the sky, running for thousands of miles. You can plan an excursion along them in a radiation-shielded rover. The radiation here is far lower than on Io or Europa, but still higher than on Earth. There is a weak magnetic field, which is unusual for a moon. It's far too feeble to ward off dangerous solar wind, though it does cause delightful little wiggles in Jupiter's magnetosphere, and faint auroras all over Ganymede's sky. There is a thin oxygen atmosphere but it's one thousand times lighter than Earth's.

It's easy to forget that the dark regions of Ganymede are made of ice and not rock. The lumpiness of the sulci suggest that rock lurks beneath the ice, which is hundreds of miles thick, and if you dig far enough down you'll unearth an underground ocean.

Ganymede isn't nearly as menacing as Io, but in the distant past it may have had ice volcanoes that gushed water, methane, or ammonia. The watery flows of these volcanoes can be just as dangerous as the hot flows you'll find on Io.

Callisto

Callisto is the one Galilean moon where it's safe to relax for a while, free from the worry of spiking radiation levels. An escape from the icy depths of Europa and hellish volcanoes of Io, it's a great place to call home while you tour Jupiter's smaller moons.

Tourist stops include the crater Tindr. Named for a god in Norse mythology, the forty-three-mile-wide depression has a messy center, where material miles below the surface rippled up during a crash from a meteor. Near the south pole is Lofn crater, one of the largest and youngest impact craters on Callisto, with a central ring 111 miles wide. Named for the Norse goddess of marriage, it's a popular place to get hitched, as long as your friends are game to travel 390 million miles for your destination wedding. Lofn is quite shallow, less than half a mile deep, and has bright rays surrounding it, which are visible from above. Its eroded rim has a gradual slope, making it ideal for a long wedding procession. You'll also want to stop by Gipul Catena, a chain of craters 290 miles long. It formed when a comet broke into pieces as it was pulled in by Jupiter's gravity.

Don't leave Callisto without spending some time at Valhalla, the largest and most magnificent crater in the solar system. It's almost twenty-five hundred miles wide, larger than the Caloris basin on Mercury, with dozens of concentric ridges radiating from the central point of impact.

Amalthea

If you want the biggest, boldest views of Jupiter available with your feet on the ground, there is no place better than the inner moons, which orbit closer to the planet than the Galilean satellites. The largest of these, Amalthea, is named after a nymph of Greek mythology. This red icy rubble pile is the third-closest moon to Jupiter. From this vantage point the planet is almost 100 times as large as Earth's full moon, big enough for you to easily admire the turbulent clouds and Great Red Spot. You'll be orbiting Jupiter in about twelve hours, but since Jupiter's average spin is a bit faster and in the same direction, it will take a while for you to get the full tour. The moon's orbit barely touches one of Jupiter's rings—the Amalthea gossamer ring, made of dust trickling off its surface. This delicate ring is ten times fainter than Jupiter's other rings, which are themselves very dim.

Leda

Counting outward from Jupiter, Leda is lucky moon number thirteen. Take a walk on this dark world, a little over six miles wide, with a surface area roughly the size of the island of Okinawa, Japan. Leda might have been a part of a large asteroid that broke apart. Though it orbits almost 7 million miles from the planet, Jupiter appears 40 percent bigger than the full moon on Earth. Over the 240 Earth days it takes to orbit Jupiter, Leda's orbit gives you views of the planet you'll never see from the more touristy Galilean moons.

Trojan Asteroids

This huge group of asteroids shadow Jupiter as it orbits the sun, preceding and trailing the behemoth planet at 60 degrees ahead and behind it. There are more than a million Trojan asteroids larger than half a mile across. They cluster in two groups, located at the fourth and fifth Jupiter-Sun Lagrange points, L4 and L5.

Although they are collectively known as the Trojans, the group at the L5 Lagrange point are the actual Trojans. A group at another point, L4, are known as the Greeks. There is an asteroid in the Greek group named after a Trojan spy, Hector. One member of the Trojan group is named after a Greek spy, Patroclus. Jupiter isn't the only planet with these rocky companions. Venus, Mars, Uranus, Neptune, and even Earth have Trojan asteroids that share their orbits at their Lagrange points with the sun.

143

Ponder THE MYSTERIES OF
SATURN

Destination ♄ Saturn

Saturn is the jewel of the solar system, home to intricately patterned rings, a kaleidoscope of colorful cloud-scapes, and a mysterious hexagon-shaped vortex. Who knew a big ball of gas could be so calming? You'll swoop through the tall, puffy clouds of its light atmosphere, where the pull of gravity is ever so slightly lower than it is back home, filling you with a familiar feeling.

Saturn caters to travelers willing to go the extra mile (or million) to get there, and to the seasoned space vacationer ready to finally see the rings of Saturn up close. Its many moons and moonlets offer diverse landscapes where you can romp or relax—on tiny clumps of ice or large rocky spheres that feel like planets themselves. You'll enjoy long walks on the beaches of Titan, one of the most intriguing moons in the solar system.

♄ AT A GLANCE

DIAMETER: More than 9 times Earth's

MASS: 95 times Earth's

COLOR: Yellowish brown with a touch of orange and occasional bluish tinges

SPEED AROUND THE SUN: 22,000 miles per hour

GRAVITATIONAL PULL: A 150-pound person weighs 137 pounds

AIR QUALITY: Thick, 96 percent hydrogen, 3 percent helium, with traces of methane, ammonia, hydrogen deuteride, and ethane

MADE OUT OF: Gas

RINGS: Yes

MOONS: 62

TEMPERATURE AT 1 BAR: -218 degrees Fahrenheit

DAY LENGTH: 10 hours and 40 minutes

YEAR LENGTH: More than 29 Earth years

AVERAGE DISTANCE FROM THE SUN: 891 million miles

DISTANCE FROM EARTH: 746 million to 1 billion miles

TRAVEL TIME: 3 Earth years for flyby

TEXT MESSAGE TO EARTH: 67 to 93 minutes

SEASONS: Similar to Earth's, but longer

WEATHER: Intermittent large storms, high winds

SUNSHINE: About 1 percent as bright as on Earth

UNIQUE FEATURES: Rings, mysterious hexagon

GOOD FOR: Skydiving, putting a ring on it, moon sports

Weather and Climate

On Earth, weather originates above us. On Saturn, it takes on a new dimension because it can storm not only above but also below. There is no ground, no dirt for rain to saturate, no trees for lightning to strike. Here it's all crowded sky with thick, billowing clouds, unrelenting winds, and planet-devouring storms. You'll take an anxious interest in how the weather shifts as you travel up and down in the sky. Like Jupiter, Saturn is made mostly of hydrogen and helium with traces of ammonia and methane. It's hazy and cold—hundreds of degrees below zero—in the topmost layers, where thick ammonia clouds seem to encase Saturn in butterscotch. As you descend, the temperature increases to one hundred below and you'll see ammonium hydrosulfide clouds, which are more reddish brown in color, like the clouds of Jupiter. Below that you'll encounter familiar Earthlike water ice clouds, just before the pressure and temperature spike to unpleasant levels.

Saturn is almost a billion miles from the sun, and a little colder than Jupiter, averaging 218 below zero at the altitude where the pressure is like Earth's. Though much less massive than Jupiter and with lower gravity, it has a pattern of eastward-flowing zones and westward-flowing belts that's similar, as is the day length, which is just ten hours and forty minutes.

It's easy to be swept away by the winds of Saturn, where speeds can reach more than 1,000 miles per hour near the equator. The year is long, almost thirty Earth years, and seasonal changes sparked by the planet's 27-degree tilt can trigger epic weather patterns. Northern spring is known for once-a-Saturnian-year storm fronts that ring the planet in turbulent flow, spreading and swirling like a drop of food coloring in water. They set off intense lightning flashes ten thousand times stronger than lightning on Earth. Do your best to avoid it, but as long as you're aboard a conductive metal aircraft, you should be safe. Like a lightning strike to a plane on Earth, the enclosed frame is an electric barrier that will protect you. The vacuum created by hot bolts of electricity moves the air fast enough to create thunder. Just like on Jupiter, thunder is not a comforting booming rumble, but a high-pitched wheeze distorted by the light gases of the Saturnian skies.

When to Go

The best time to visit Saturn is when Jupiter is available to provide a powerful gravity assist. Alignments happen roughly every twenty years. Check with your local Intergalactic Travel Bureau office for the next available window.

Spring is a wonderful time to visit Saturn, when the hexagon at its north pole sharpens after a seven-year winter. The slightly more intense gaze of the sun spurs Saturn's gases into motion, triggering beautiful and turbulent storms.

Getting There

If you want to get to Saturn in a hurry, and the thought of a ship powered by one thousand nuclear bombs doesn't frighten you, you might consider reviving Project Orion, a NASA program from the 1960s. If everything goes smoothly, the level of radiation you receive from outer space will exceed any you might be exposed to from your nuclear

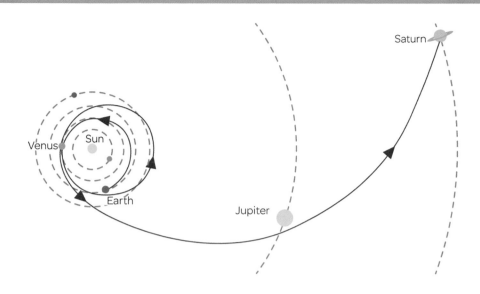

Inspired by Cassini spacecraft, this example flight path uses a gravity assist from Jupiter to help you get to Saturn before your vacation days run out.

bomb-powered ship. If you choose this option, you'll launch from Earth in a chemical rocket and rendezvous with a special ship in space, to avoid exposing Earth dwellers to fallout from the blast when it takes off.

If you prefer to ride a more traditional rocket all the way to Saturn, you can use gravity assists to complete the 794-million-mile trip. Earth, Venus, or Jupiter can give you a boost. While it seems counterintuitive to get a gravity assist from an inner planet to get to an outer planet, the extra boost may more than make up for the mileage, and provide good opportunities for scenic photos.

149

When You Arrive

One of the first things you'll notice as you approach the tan gas planet is the beauty of its rings. From a distance these rings seem solid, flat, and static. As you get closer, their solid appearance resolves into individual fragments. Some of these floating space icebergs are larger than buildings, and others are microscopic. They circle Saturn under the influence of gravity, with closer bits orbiting faster than the ones farther out. In some of the large hollow gaps between the rings you might see a small moon.

You'll also notice its flattened ball shape. While no planet is a perfect sphere, Saturn shows even more of a central bulge than average. It's 10 percent wider at the equator than it is at the poles (Earth's is a fraction of a percent wider), as a result of its rapid spinning.

As you approach you may find it hard to believe the clouds aren't an optical illusion, a watercolor masterpiece suspended in space. The colors are caused by trace amounts of methane, ammonia, and other gases, organized in striped bands and zones across Saturn's face. They get wider closer to the equator, with each stripe representing a distinct jet stream in the planet's atmosphere. Drawing nearer, you'll see swirling teardrop-shaped storms.

You'll probably enjoy a jaunt through its atmosphere to get a closer look at its wispy clouds, as varied in shape and size as on Earth. The top layer is made of ammonia, bright with a muted yellowish-brown tint caused by traces of sulfur in the clouds. These traces are a kind of natural smog, an innate feature of the planet present billions of years before humans would emerge on Earth to create smog-bellowing cars.

Getting Around

The skies of Saturn are filled with 96 percent hydrogen, the lightest element in the universe. If it were possible to put the planet in a gigantic bathtub, it would float because it's less dense than water. This poses a unique engineering challenge for airship travel. On Earth, hydrogen can lift large balloons and airships because it's so much lighter than our (relatively) heavy air. On Saturn, the only way to get airborne is to use heated hydrogen, which is less dense than the colder surrounding hydrogen. Sound dangerous? Don't worry—as long as you're careful not to mix it with oxygen near a flame it won't catch fire. Vacuum airships are another option for touring the skies

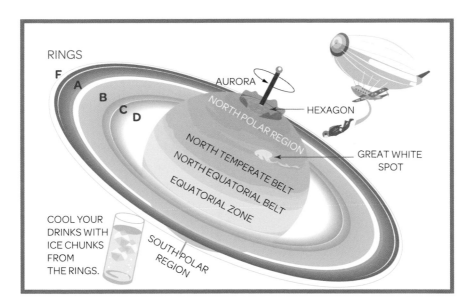

Ring in the new year on Saturn.

of Saturn. After all, the only thing that's lighter than hydrogen is . . . nothing. Unfortunately, vacuum airships have a tendency to implode.

Saturn is a marvel from orbit. Though the atmosphere is light, the planet is still massive; you'll use three times more rocket fuel escaping Saturn's gravity as you did to escape Earth's pull 794 million miles away. You'll catch a shuttle to visit Saturn's surrounding moons, and tour most of them by rover. The biggest moon, Titan, has a superthick, nitrogen-based atmosphere—ideal conditions for small airships and airplanes.

What to See

North Polar Hexagon

Over the north pole of Saturn you can view one of its most enchanting and mysterious features: the hexagon. Also called the northern polar

vortex, this whirlpool of swirling gas forms a slightly rounded but very distinct six-sided geometric shape. At more than two and a half

The hexagon on the north pole of Saturn is formed from a jet stream–like wind whose waves curve to form the shape.

NASA/JPL/SPACE SCIENCE INSTITUTE

times the diameter of Earth, sixty miles deep, and with each edge longer than Earth is wide, it's one of the greatest natural wonders of the solar system. If you descend into the hexagon, be prepared for some serious turbulence, as winds exceed 220 miles per hour. The whole hexagon rotates once about every ten and a half hours.

What could possibly create such an unusual phenomenon? Saturn's winds flow at different speeds as you move deeper into the surface and along the surface itself. These winds create giant waves. With six peaks folded around the pole of Saturn, these giant waves form a hexagon. The number of peaks in the wave—and the shape you end up with—depends on the speeds and the differences in the speed of different layers of wind.

Southern Hurricane

The south pole of Saturn may not have a hexagon, but it does have a never-ending hurricane that looks like a giant unblinking eye. Winds there are treacherous, traveling at 350 miles per hour—faster than Earth's hurricanes. Adventurous travelers may wish to descend deep into the eye. It is one of the few places other than the hexagon where you can dive a little deeper into the atmosphere without being engulfed by clouds.

Activities

Surf the Rings

Many humans have glimpsed through telescopes Saturn's majestic rings, but few have made the 750-million-mile journey to touch them.

Experiencing the rings of Saturn is a lifelong dream for many an aspiring space tourist. People can spend weeks meditating on the intricate patterns.

Saturn has seven main rings, each made of hundreds of thousands of narrower rings called ringlets. They are gray, with a hint of shimmering pink and brown, and each is named after a letter of the alphabet. The rings spread from forty thousand to three hundred thousand miles from the planet. You could fit half of Saturn's width in the space between the top of Saturn's clouds and the inner edge of the rings.

The closest ring to Saturn is the D ring. The C and B rings (B is the beefiest and brightest) are farther out, respectively. A large, famous gap called the Cassini Division separates these inner rings from the rest. Contrary to popular lore, the Cassini Division isn't completely empty, but contains a scant collection of dust.

The next ring beyond the Cassini Division is the A ring, which is home to the smaller Encke Gap. This is followed by the F ring. The outermost rings are the tenuous G and E rings.

The rings are extremely thin, only about thirty feet thick, so it's best to view them from above or below rather than from the side, where they look like a sliver. As you get even closer you'll see the particles aren't as organized as you might think. They clump into irregular curves, elongated in the direction of motion, and there is plenty of space between the clumps. If you could collect the contents of all the rings together, you'd have an entirely new, medium-size Saturn moon.

The rings are not static, but shrink and grow as if they are alive. Saturn's moons affect the motion of the rings, gracefully playing with them, making them ripple and dance with their gravitational influence. The rings can flare and become thicker, forming jagged, mountain-like shapes on the edge. Dark spoke patterns cross the rings, appearing and then disappearing. The rings are fed by ice spewed up by moons and incoming comets. There is probably less material coming in than leaving the rings. Over a long time, they may drift away entirely.

Though no one knows for sure the origin of Saturn's rings, the best theory suggests they were formed by a moon that got too close to the planet. The difference in gravitational pull on the close side compared to the far side caused the moon to stretch out until it was shredded by these tidal forces.

A snapshot showing the backlit rings of Saturn, looking back toward the sun.

NASA/JPL/SPACE SCIENCE INSTITUTE

Enjoy a space cocktail on the rocks with ice directly from the rings, which are about 93 percent water ice. Make sure your bartender avoids using ice tainted with pebbles, salt, or toxic chemicals.

Look for Auroras

While you're visiting either of Saturn's poles, be on the lookout for auroras. Like on Jupiter, Saturn's magnetic field captures charged particles from the sun, its moons, and the planet itself, sweeping them into the magnetic poles. When the energetic particles hit the gases of Saturn's upper atmosphere, they give energy to the atoms and molecules, which release it in the form of light.

While Saturn's magnetic field isn't nearly as strong as its neighbor Jupiter's, its lights are just as beautiful. The auroras look like huge curtains swaying high in Saturn's atmosphere at a height of sixty miles. The auroras change color with altitude—pink on the bottom, fading to a lovely violet hue higher up.

Storm Chase

Saturn's churning atmosphere looks like a well-swirled bowl of coffee ice cream and chocolate sauce. Every few decades, a giant storm nearly encircles the planet, to the delight of storm chasers around the solar system. These Great White Spots are a rare sight, with only six happening over the past 140 Earth years, or about once every northern Saturnian summer. Larger in size than the United States, they are thought to be formed from rising and sinking water vapor in the atmosphere, similar to the basic mechanism that causes storms on Earth but taking much longer to develop.

Skydive

Saturn is more massive and larger than Earth, providing unique conditions for skydiving. It's important to choose a reputable company, and rule number one is to book with someone who takes payment after you have safely completed your fall. Disreputable operators have a habit of starting their skydivers too high, where they accelerate too quickly. By the time these unlucky adventurers reach the denser gases they are traveling so fast that they burn up like meteors. Don't let this scare you; as long as you start well within the atmosphere, the drag will prevent you from accelerating to a fatal speed. Starting out where the pressure is the same as Earth's, you feel just a few pounds lighter as you stand at the open doorway of the airship, ready to jump out. Once you take the leap, your acceleration is only a little slower than it would be on Earth. Instead of slowing down once you've reached 120 miles per hour, the terminal velocity on Earth, you'll make it up to 320 miles per hour before your speed levels out; Saturn's atmosphere is less dense and causes less drag. You can enjoy the exhilaration of traveling faster than the fastest race car with only a space suit separating you from the open sky.

You'll start your fall in a layer of wispy, yellowish ammonia ice clouds. After falling ten minutes or so you'll have traveled more than sixty miles, and you'll start to encounter thicker, redder ammonia hydrosulfide ice clouds. Finally, you'll reach some more familiar, white water vapor clouds. The entire fall will be very dark—the sunlight is only 1 percent that on Earth at the cloud tops, and rapidly gets darker the deeper into the atmosphere you fall. Eventually the skies will fade to pitch-black. Feel free to fall for a bit into the dark abyss. You're in no danger of hitting the ground because there is no ground to hit, but it is possible to go so deep that your suit implodes under the pressure. You'll probably get bored of falling in the darkness, activate your parachute, and fire your ascent rockets to rendezvous with your recovery ship before that happens.

If you were to keep falling far beyond where you could survive, you'd encounter the most mysterious region of Saturn, an area where solid diamond condenses out of the high-pressure and -temperature slush. The rough diamonds can reach sizes larger than an inch and they fall as diamond rain. While all the gas and ice giant planets hide diamonds in their interior, Saturn is thought to conceal the greatest amount of this enchanting precipitation.

See the Fireworks

Saturn's atmosphere is about 96 percent hydrogen, which, in the presence of oxygen, becomes very explosive. While wasting precious oxygen is frowned upon, and fires inside habitats and spacecraft are dangerous, on special occasions it's possible to collect hydrogen from the atmosphere to ignite it in a controlled explosion within an artificial-air environment. The fiery plumes from the ignited gas look like mini mushroom clouds.

What's Nearby

Why settle for a single moon? Saturn has sixty-two satellites, which host a variety of fun for the knowing moon-hopping traveler. However, if you only have time to visit one of Saturn's moons, make it Titan.

Titan

If you're looking for a beach vacation anywhere other than Earth, Titan is your best option. Be warned—a visit to Titan will feel more like an expedition to the Antarctic than a sun-filled getaway to Cancún. The beaches take some getting used to, as they are cast in a perpetual dusk. High noon feels like fifteen minutes after sunset, and the sun is obscured by an impenetrable layer of haze and clouds. If you can look past that, the rocky beaches of this orange moon are quite lovely.

You'll need protection from the toxic chemicals in the air and subarctic temperatures, but you can ditch your pressurized suit since the atmospheric pressure is just one and a half times what it is at Earth's sea level. You'll feel the weight of it pushing on you as your clothes clings to your skin. It's about what you'd feel at the deep end of a diving pool.

It's fun to wade in the hydrocarbon lakes and rivers. Since the hydrocarbons are heavier than water, while wearing a thermal suit you can float high on the viscous liquid. Try dolphin-like jumps out of the water. If you are still and patient, you may hear the sound

Saturn's moon Titan is one of the most fun places in the solar system.

NASA/JPL/SPACE SCIENCE INSTITUTE

159

of crashing waves from the generally calm seas. This sound is low and unfamiliar, distorted by the alien atmosphere and low temperatures. The waves travel slower than they do on Earth, owing to the thick liquid, thick air, and slow winds.

As you spend time on the beach, you'll probably notice the proliferation of Frisbee games. Titan's atmosphere makes the moon particularly suitable for throwing a disk, although even experienced ultimate players must take time to adjust their game. The super-dense atmosphere leads to a lot of lift, a slow downward fall, and a lot of air resistance. You'll need a strong throwing arm.

Along with beachgoers, the numerous methane and ethane lakes draw droves of boaters enjoying the views of the shores and the haze above. First-timers may be surprised by how low boats sit on the lakes and rivers, due to the low density of methane. Sailing is easy through the low viscosity seas. A popular spot for boating is Kraken Sea. It's the largest body of liquid on Titan and is just larger than Earth's Caspian Sea, and gets as deep as 656 feet in places.

Surfers also crowd this moon because of its reputation for strange swells. Don't be duped; the moon has calm winds and smooth seas in winter. Waves are typically small, only a couple of inches tall and slow moving at about 1.5 miles per hour. If you're lucky enough to catch a rare big wave, surfing Titan is an othermoonly experience. You might find good surf when there are hurricanes brewing over the polar seas.

Storms can bring methane showers, with huge drops almost twice as big as Earth raindrops. Rain falls slowly on Titan, like snow-flakes on Earth, due to the thick atmosphere and low gravity. You might spot flashes of lightning and hear alien thunder during a storm.

After you've had enough of Titan's beaches, head to the

dunes at the equator. With sand the consistency of granola and as dark as asphalt, they have an austere beauty. You can explore them in a rover, stopping to go dune boarding when you find some steep slopes. If you get the chance, try to view the dunes from the air. They encircle Titan at the equator.

One of the best ways to explore the moon is by flying under your own power. The soupy thickness of the atmosphere and low gravity mean that with a wing suit, a flap of your arms, and perhaps a slight thruster assist, you'll be airborne like a bird. Moving your glider-like wings in thick air is difficult, even in the low gravity. Drops of gasoline dew might condense on your faceplate. To get airborne, run as fast as you can, and then engage your assist thruster and jump as your feet lift off. Once your feet leave the surface, flap your wings as hard as you can and you'll rise to the sky. When the ground has nearly disappeared into the thick orange-yellow haze below you, try diving down to skim the surface of one of the methane lakes.

If history is your thing, you can visit the remains of the Huygens probe in a region called Xanadu, a bright area of the southern hemisphere. It was the first probe to land anywhere in the outer solar system, taking the first snapshots of Titan's surface, which showed a rocky, desolate terrain.

Pan

Get a close-up view of Saturn's rings from the surface of walnut-shaped Pan, the closest moon to Saturn. It gathers up particles that would other-wise make a ring around Saturn, forming the Encke Gap. In Greek mythology, Pan is the god of shepherds, and in this case Pan uses gravity to shepherd ring particles.

Pandora

Pandora is bright and icy. We recommend you find a nice hollow pit—Pandora is full of them—and hunker down inside. Then you can watch Saturn as you orbit around. Because Pandora rotates at the same rate as it orbits, in synchronous rotation, you'll keep that nice view of Saturn as you loop around it.

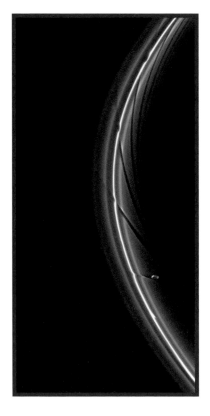

Prometheus

Shaped like a potato, this icy space spud is eighty-five miles long. Prometheus may be small, but it sculpts and shapes its nearby ring, the F ring, considered Saturn's weirdest ring. Prometheus's gravity creates lasting ripples, breaks, and kinks in the ring. Ride this moon for a front-row seat to the complex and hypnotizing dynamics of the F ring.

The moon Prometheus makes waves from its gravity in Saturn's rings.

CASSINI IMAGING TEAM/SSI/JPL/ESA/NASA

Daphnis

Daphnis orbits within the Keeler Gap, in the A ring, clearing it of particles. It is one of two moons (the other being Pan) that orbit within the main body of the rings of Saturn. Riding on Daphnis you'll get an unparalleled close-up view of the rings, revealing the icy boulders, pebbles, and tiny flecks that make up their intricate rippling patterns. You'll also see mile-high waves on the edge of the gap, which cast long shadows onto the rings.

Mimas

Star Wars fans may want to travel the 800 million miles to Saturn just to see Mimas, a 120-mile-wide moon that looks exactly like the Death Star. A giant asteroid left a huge gash in its side, leaving the crater Herschel. The impact responsible for this crater must have nearly destroyed the moon, leaving an eighty-mile scar that gives it its menacing look.

Mimas is also responsible for the Cassini Division, although its orbit

The prominent crater Herschel makes the moon Mimas look like the Death Star.

NASA/JPL/SPACE SCIENCE INSTITUTE

163

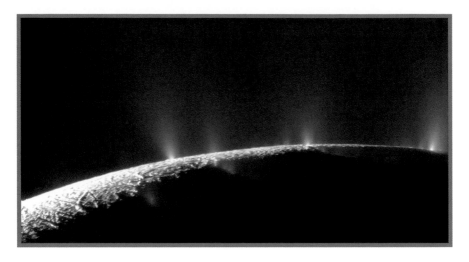

This ice moon's geysers shoot water from the cracks in its surface, some even escaping the moon's gravity and becoming ice particles in Saturn's E ring.

NASA/JPL/SPACE SCIENCE INSTITUTE

is far outside it. Particles at the edge of the Cassini Division orbit in exactly twice the time it takes Mimas, which is farther out, to orbit Saturn once. This means that the particles tend to line up regularly, and each time they line up, they experience a little extra gravitational tug. Over time, these extra tugs moved the ring particles out of the Cassini Division and left a gap.

Enceladus

A bright and shiny moon, Enceladus is the ice king of the Saturnian satellite system. Despite being only three hundred miles across, one-seventh the size of Earth's moon, its surface is varied, with giant chasms and crater fields. While visitors must contend with rugged terrain, the moon's vistas will make you and your camera happy.

Enceladus is extra cold because it reflects so much light. As cold as it is, it is not frozen solid. The inside is warmed by tidal heating—the gravitational squeezing and stretching effects of Saturn's uneven gravity. The heat melts a liquid water ocean layer under its thick blanket of ice.

Because of the moon's stretching and compressing, you'll be able to tour huge tectonic fractures called tiger stripes in its brittle, icy shell. Each stripe is a chasm, called a sulcus, about eighty miles long, more than a mile wide, and a third of a mile deep. They are common around the south pole of Enceladus, and you can hike the sulci, walking from the easternmost stripe, Alexandria, to Cairo, to Baghdad, and finally to Damascus. The area around the stripes, while still quite chilly, is warmer than other parts of the moon, and a hot spot for the moon's spectacular main attraction: geysers.

Enceladus's geysers—and the plumes they create—are another one of the natural wonders of the solar system. There are more than a hundred of them, lining up along cracks in the surface, spraying frequently. Standing near a geyser, you'll see a plume jutting far into the atmosphere. The spray will tower over you, rising higher than a hundred miles, shooting out water at an impressive 800 miles per hour. The water and vapor quickly freeze, and you will be blanketed in a gentle rain of sparkling ice crystals. Because there is no atmosphere to interfere, the salt-rich crystals arc into a large umbrella shape as they fall down, like the lava from the volcanoes on Jupiter's moon Io. Some freshwater ice is launched off of Enceladus, continually resupplying glittering ice particles to Saturn's outermost ring, the E ring.

After you check out the geysers, take note of the Snowman craters—three giant craters that line up to look just like a snowman. These craters are filled with thin cracks, which appear across much of the moon.

Hyperion

This irregular-shaped moon is full of holes and craters, and looks just like a sponge. You'll never know what the view from the surface will hold next because this moon tumbles, rotating unpredictably as it orbits Saturn every twenty-one days. Be careful not to fall into any of its deep pits as you walk, or you might find yourself wandering lost in its maze of icy underground caverns.

Iapetus looks like it's splattered with paint on one side.

CASSINI IMAGING TEAM/SSI/ JPL/ESA/NASA

Iapetus

The third-largest moon of Saturn, the ice world of Iapetus looks as though someone dropped a giant can of dark paint on one side. Its leading hemisphere is brownish black with splattered edges, while its trailing hemisphere is bright. Iapetus has a tall ridge that runs three-quarters of the way around it at the equator. Mountains on the ridge rise twelve miles above the surroundings, over twice the height of Everest. The moon's very low gravity, less than a quarter of Earth's, makes climbing the steep slopes a little easier. Standing on top of them will leave you breathless, particularly if you can see Saturn from your vantage point. Iapetus is the best moon for viewing the rings because it is the only one whose view is not edge on, and Saturn appears four times the size of the full moon in Earth's sky.

Destination 🜨 Uranus

Uranus is the secret gem of the solar system. It's a planet known for its kooky magnetic field, epically long seasons, flailing inner moons, and its most distinct characteristic: its sideways tilt. Compared with Earth and all the other planets in the solar system, it is toppled over, rotating along its orbit at 98 degrees. You'll hardly notice this quirk, since gravity keeps you securely docked in the planet's gaseous skies and north is still defined as the tip of this overturned top. No one knows for sure why Uranus roams around all askew, though odds favor a savage collision with a celestial rogue about the size of Earth during the formation of the solar system.

Your bravery in embarking on the long journey to Uranus will pay off, when you get to know a truly one-of-a-kind vacation destination. Once you arrive, there is plenty to entertain and delight you. Bask in the blue-green glow of this icy giant's gaseous methane skies from one of its aerial cities, floating at an altitude where the pressure and gravity will remind you of home.

Visitors also enjoy its many moons—with crevices and chasms that embarrass the Grand Canyon. Its majestic primary satellites, named after characters dreamed up by William Shakespeare and Alexander Pope—Umbriel, Miranda, Ariel, Titania, and Oberon—contain numerous amusements within their chains of craters and trenches.

AT A GLANCE

DIAMETER: 4 times Earth's

MASS: 14.5 times Earth's

COLOR: Pale blue

SPEED AROUND THE SUN: About 15,000 miles per hour

GRAVITATIONAL PULL: A 150-pound person weighs 133 pounds

AIR QUALITY: Thick, 83 percent hydrogen, 15 percent helium, 2 percent methane

MADE OUT OF: Gas

RINGS: Yes

MOONS: 27

TEMPERATURE AT 1 BAR: -323 degrees Fahrenheit

DAY LENGTH: 17 hours and 14 minutes

YEAR LENGTH: 84 Earth years

AVERAGE DISTANCE FROM THE SUN: 1.8 billion miles

DISTANCE FROM EARTH: 1.61 billion to 1.96 billion miles

TRAVEL TIME: 9 Earth years for flyby

TEXT MESSAGE TO EARTH: 144 to 175 minutes

SEASONS: Long

WEATHER: Beautiful aurora and lightning, methane haze covers storms in cloud layers

SUNSHINE: Eternal nautical twilight

UNIQUE FEATURES: Toppled over

GOOD FOR: Moon sports, bungee jumping

Weather and Climate

Don't let those majestic blue images of Uranus fool you; the weather on this planet is the strangest in the solar system. Each season lasts twenty-one years—bringing a whole new meaning to the phrase "Winter is coming"—and the planet's extreme tilt makes for some odd patterns. Its south pole faces the sun for half of its eighty-four-Earth-year-long year, meaning the sun doesn't set for 15,340 Earth days in that region, even though the local day lasts 17 hours and 14 minutes. During southern summer, the sky seems as calm as a lake just before dawn. Of course, even in summer, this lake will be frozen. And it's not exactly a lake, but a ball of churning ice and gas made from hydrogen, helium, and methane. When seasons change, it's not unusual for dark spots—storms two thousand miles across—to appear, accompanied by wispy, bright white methane clouds that space meteorologists refer to as bright companions.

To the naked eye, Uranus looks like a serene blue sphere.

NASA/JPL-CALTECH

The weather wavers, according to season, between peace-fully cold and stormily cold. If you love a good old-fashioned arctic blast, you'll find Uranus's temperatures quite suitable. It's nearly 2 billion miles from the sun and has the planetary record for coldest

171

atmosphere in the solar system—even colder than Neptune's at times. In the heat wave of a twenty-one-Earth-year-long summer, you'll never see temperatures top -300 degrees. That's more than twice as cold as Earth's record low, recorded at Vostok Station in the Antarctic. When the sun is one four-hundredth as bright as it is on Earth, there's just nothing that will ever warm you.

Don't forget to pack your Windbreaker. High winds are the norm, topping 560 miles per hour. Storms the size of the United States, at twice the power of a Category 5 hurricane on Earth, rage regularly. The clouds host electrical storms with lightning flashes, and auroras appear during periods of high solar activity.

When to Go

Since the year is so long, technically you could visit Uranus for many Earth years and see only a few Uranian months go by. High tourist season arrives every forty-two Earth years, during the solstice, when one pole faces the sun and the other is completely cast in shadow. Imagine the feeling of seeing the sun set for the last time in your life, in middle age. That's what it's like in the skies opposite the sun as solstice approaches.

If you prefer to experience decades-long midsummer daylight, plan to arrive at the sun-facing pole soon after equinox, in plenty of time to prepare to celebrate Uranus's solstice around 2028. At that time, day and night will be distinguished only near the planet's equator. Due to Uranus's quick rotation those days are like short winter days near the Arctic Circle on Earth, but let's be honest—when you're 1.8 billion miles from the sun there isn't much of a difference between day and night.

If you want to save fuel, it's best to go to Uranus when Saturn or Jupiter are available for a gravity assist. Don't forget to allow about a decade years' travel time each way. Storm chasers should plan to arrive in 2049, when there is plenty of storm activity.

Getting There

If you plan it just right, you can make the 1.8-billion-mile trip to Uranus in less than a decade. Consider how much vacation time you want to use, whether you can work or continue your education while en route, and if you'd like to return.

Nuclear bomb–powered rockets will be your fastest option, and they offer the possibility of a return trip. The risk may be worth it, if you want a chance to see your loved ones back on Earth before you grow too old to travel home.

Cryogenic preservation is a handy option for those who get bored on long trips. Vitrification, a process that preserves tissue at low temperatures without damage, has been used in organ transplant procedures successfully, though full-body recovery is . . . questionable. The technology hasn't been perfected, so watch out for companies offering guaranteed rehabilitation. Read the fine print, as the warranties on revival might not take effect for another fifty to one hundred years. If it does work, you'll be able to make the trip without shortening your life span (too much). You may choose to volunteer for a research program in exchange for space fare. Decide what's best for you.

For those who wish to stay conscious and active during the journey, your ship will offer similar amenities to those on a cruise ship, but with a lot less personal space. You will be able to keep up

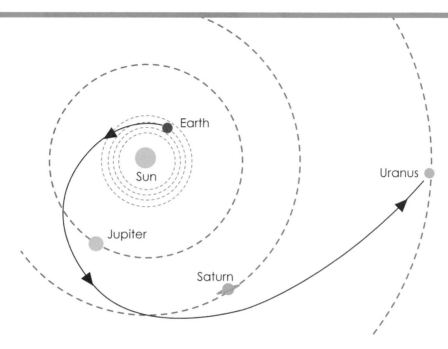

As you venture to the edges of the solar system, you may consider going into hibernation during your journey.

with news from Earth and the latest episode of your favorite shows. Just remember that you'll have to deal with longer communication delays as you travel away from Earth.

When You Arrive

So you've made it to Uranus! Whether you've aged a decade, traveled on a retro-style atomic rocket, or survived the vitrification process and woken up in the future, it's time to celebrate—and set your clock.

Though the Uranian day is just seventeen and a quarter hours long, visitors keep Earth time because, well . . . You know the feeling of being in the middle of a calm sea in the evening, just after that moment when the horizon has melted into invisibility and the outlines of objects are only vaguely detectable? That's the brightest it gets on Uranus. Ever. Call it eternal nautical twilight. Not to worry—you can get your daily dose of vitamin D at a variety of tanning salons and sunrooms located throughout the planet; they're as ubiquitous as Starbucks are on Earth.

Far from it, Uranus often appears as a giant featureless blue ball. Only when you're very close can you see that the planet is an endless ocean of thick blue fog, gently undulating and stretching in all directions to the horizon.

Once you've reached Uranus, it takes up to three hours (2.39 to 2.93 hours, to be exact) to send or receive messages with your friends and family back at home. It's all #latergrams on Uranus.

Getting Around

When you're done admiring Uranus from orbit, you can drop down into its gusty skies to get a closer look at its blue-tinged clouds. There is no solid ground, so you'll stay in aerial cities.

Airships are great for leisurely rides through the helium and hydrogen atmosphere. For faster trips you can ride aboard an airplane. The main jet stream travels westward at the equator, against the planet's rotation.

When you've had enough of the hazy skies, you'll have to air-launch back into orbit. While touring Uranus's varied moons, you can rent a rover or hopper.

What to See

Equatorial Waves

At first glance Uranus may seem as flat and boring as an endless cornfield, but things get interesting if you take some time to look below

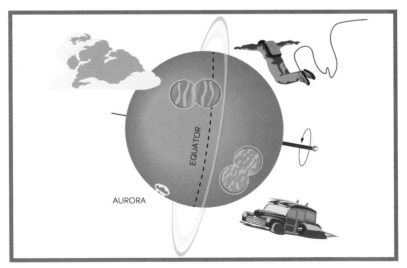

There is plenty to do and see within this mesmerizing blue orb.

the surface. One of the strangest attractions of this planet, a massive weather pattern that encircles the planet south of its equator, can only be seen in infrared light. You can see this cloud pattern with special goggles that pick up long-wavelength light, at energies that are invisible to the naked eye. The goggles amplify the light so that you can see more detail, much like night-vision goggles do. Only then will these scalloped atmospheric waves become visible, a beautiful sinuous pattern for your enjoyment. Observe it from a safe distance, or take your ship once more into the breach.

A view in infrared light reveals waves and spots hiding behind Uranus's often-calm outer haze.

NASA/ESA/L. A. SROMOVSKY/P. M. FRY/H. B. HAMMEL/I. DE PATER/K. A. RAGES

The Turbulent North Pole

After you've had your fill of trippy waves near the equator, work your way north past some stormy hot spots to the swirling north pole. Much like Jupiter's and Saturn's north poles, this area is full of eddies, with high clouds mixed with atmospheric holes, giving it a lovely dappled appearance and possibly a wild ride. As with all planetary weather patterns, keep an eye on the forecast. Even storms that have been around for many, many years can suddenly dissipate, and new ones may appear at any time.

The Rings

Uranus has thirteen rings, made up of orbiting objects ranging from the size of a piece of sand to a small sofa. The rings are ghostly and darker than charcoal. A popular pastime is to float with the rocks. Enjoy the silent solitude of your shared orbit as you admire the methane haze below. You are now a part of the rings.

Diamond Lakes

What could be more alluring than a lake of liquid diamond, with precious stones floating on the surface like icebergs? Such a magical place is thought to exist deep in the interior of each of the gas planets. However, don't be tempted by the siren call of these luxurious lakes. You will certainly perish in your futile attempt to reach them, crushed and melted by the immense pressures and temperatures.

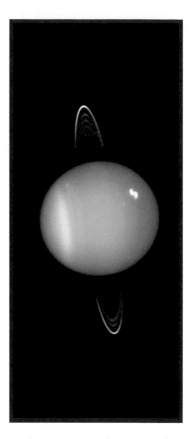

NASA/ESA/M. SHOWALTER (SETI INSTITUTE)

Activities

Tour the "Venice of the Sky"

Always dreamed of sleeping among the clouds? The buoyant airships of Uranus, tethered together in a miracle of urban architecture navigable by gondola balloons in the spirit of Venetian canals, promise to make your dreams come true. These floating cities drift with the winds around the globe once per Uranian day, and migrate from the north pole to the south pole from solstice to solstice, following the meager sunshine. Air pressure at cruising altitude is comparable to sea level on Earth. Homesick for Earthlike gravity produced naturally by the presence of a planet? That's close to what you'll feel amid Uranus's methane clouds.

Not everything will remind you of home. Like its cousins Saturn and Jupiter, Uranus has a chilly gaseous upper layer of hydrogen mixed with helium and methane. The atmospheric gases soak up all the red light, lending Uranus its iconic blue hue. The lower layers reveal a planet of a fluid nature with an icy mixture of water, ammonia, and methane.

While the hydrogen-laden atmosphere poses some safety risks because of the explosive nature of hydrogen-oxygen mixtures, it provides a source of fuel for use in the ascent out of Uranus's atmosphere and for side trips to the moons.

There are plenty of ways to enjoy the many delights of these lofty cities. Suit up and take a romantic walk along one of the famous piazzas to gaze upon Uranus's rapturous blue vistas. No need to bring a pressurized suit on your evening stroll—grab a thermal suit and a breathing pack and take in the view. Watch for lightning in the clouds below and bask in a fascinating type of aurora known as electroglow, viewable near the sunlit pole and produced by ultraviolet energy there.

Unwind at a Heliox Club

An abundant supply of helium feeds a vibrant nightlife throughout the aerial cities. As with oxygen bars on Earth, customers can choose from a wide variety of flavored mixes of oxygen and helium. An evening of silly giggling made possible by helium inhalation is a fun and relaxing way to unwind after a day of balloon tours. Nightly high-pitched karaoke contests fill up quickly and competition is fierce. There are helium-free options available at most heliox clubs.

Take an Airship Tour

Want to explore the hazy mysteries of the Uranian atmosphere? Lose yourself in the pale blue fog with a dirigible tour. These specially designed vessels look like a cross between a submarine and a hot air balloon, and are used to navigate the planet's odd atmosphere. Most aerial cities have ports filled with them where local tour guides will compete for your business. You can traverse the inner layers during a daylong outing in your own private capsule, or enjoy one of the many group tours to get a closer look at the skies.

The serene and constant blue planet will reveal amazing detail as you descend. At first it will be like you're surrounded in blue sky. The blue color is from the same scattering of light that causes Earth's blue sky, combined with methane blocking the red light. You may find your Earth-adapted mind playing tricks on you as you perceive a hazy but solid ground below you. Only as you gradually break through the thick effluvium and the light from above dims will you realize there is truly no solid ground below. Eventually the endless sea of blue fluid haze will break into a layer of white methane clouds. After lightning flashes, you'll hear the strangely altered sound of Uranian thunder, the pitch changing with temperature and density of the air. Below the white methane clouds you'll pass several additional layers. Yellow ammonia clouds are followed by the ruddy ammonium hydrosulfide clouds. Not to worry, your pressurized capsule will protect you from the noxious smell of rotten eggs (hopefully). Beyond these smelly clouds are the familiar white water ice clouds. Below you there is only a deeper, darker, thicker ocean of murk.

Though your ship will surely implode before you reach it, the pressurized sea of water, ammonia, and methane gives way to an expanse of liquid metal, like quicksilver but made of hydrogen. Below that are ten Earths' worth of molten rock. Even if your capsule could withstand the extreme pressures and temperatures of the depths of Uranus, you'll need to turn back, or risk being trapped forever by its high gravity.

Jump into the Abyss

Strap on a tether and jump from one of the many adventure stations littered throughout the city. You're usually required to pay in advance, in the event you have second thoughts once you understand just what it means to jump headfirst into the void. Stepping across the platform is not unlike walking the plank of a ship, except instead

of shoving off into freezing salt water, you'll drop into a veil of gaseous haze. Tour groups will wait expectantly as they watch you weigh the possibility of backing out. Then once you jump, time stops and it's just you and the god of the sky.

What's Nearby

Uranus has more than two dozen moons, with names that fans of William Shakespeare and Alexander Pope will surely enjoy. Each has its own charms, and the smallest of the twenty-seven is just ninety miles in diameter. Whether you're looking for a weekend excursion or many months of moon-hopping fun, there's a satellite for every taste.

Uranus's moons are home to some of the most avid sports enthusiasts in the solar system. Expect to find moon-class facilities with plenty of ice-skating, hockey, tennis, volleyball, golf, and rock climbing.

Consider a weekend trek to an inner moon, like Cordelia, Ophelia, or Mab, which offer front-row views of the Uranian ring system and a bit of rousing danger. Rumor has it a moon named Desdemona is set to collide with Cressida or Juliet within the next 100 million years.

Titania

The fairy queen Titania is also fair queen of the Uranian moons, as she is the biggest and most massive for 100 million miles, and claims the title of eighth-largest moon in the solar system.

A favorite tourist spot on her cratered shape is Messina canyon (Messina Chasma). At more than 932 miles in length, it's twice as long and many times wider than the Grand Canyon. After a long day of exploring its rim by rover, relax in the evening at one of the many ice hotels that cater to families and couples alike. Indulge in an invigorating sauna followed by a carbon dioxide ice bath. Before bed, read your favorite passages from *A Midsummer Night's Dream* and puzzle at the fact that you, humble traveler, will sleep nestled in the care of a fairy queen.

Looking for a bit of a thrill? You'd probably like to dive off of one of Titania's shockingly high cliffs. One towers two and a half miles above the surface. Bungee jump in its low gravity, where the average person won't weigh more than about ten pounds. It's sure to be a leap you won't soon forget. Though experts characterize Titania's surface as fluffy, make sure your rope is nice and secure, because even at low gravity, if you fall long enough you'll catch enough speed to hurt yourself. After a two-and-a-half-mile fall on Titania, you'd hit the ground with a speed of 120 miles per hour.

Be sure not to miss the north end, home to this moon's most recognizable craters. Gertrude, named after Hamlet's mother, is 186 miles across. Nearby Ursula is about half the size, and Calphurnia, to the west, contains an odd ellipse-shaped peak. Notice a naming trend? They're all female characters from Shakespeare's plays. The practice is inspired by John Herschel, the son of the discoverer of Uranus, who was first responsible for naming its moons.

Oberon

If craters named after Shakespearean characters fascinate you, hop on over to Oberon. King of the fairies, he's the husband of Titania, and saturated with craters. Old canyons and craters on top of more craters distinguish him, as do hints of significant geologic activity in his past. The craters are numerous but shallow, one of the largest being the famous Hamlet, which is 124 miles across. Othello is there too, in crater form and fifty-six miles wide. The Mommur canyon (Mommur Chasma), Oberon's largest, is shorter than Messina canyon on Titania, though about three times as deep.

Mountain climbers will love an as-yet-unnamed peak in the southeastern region, a mile and a quarter higher than our own Mount Everest. You could be the first to climb it—and claim naming rights in the process.

Like many of the other moons nearby, Oberon has plenty of water ice on the surface (though it lacks dry ice), and temperatures range from -300 Fahrenheit in the summer to a frigid -400 in winter.

Miranda

Do you have a craving for the bizarre? Meet Miranda. Miranda is a Humpty Dumpty moon, porous and gray, and seems to have been blown apart in the distant past, only to be put back together again. She is the most mysterious moon in the solar system, with a surface even the most dedicated researchers struggle to explain. She's got cliffs, faults, ridges, craters, scarps, and canyons. She's got short canyons, long canyons, and a canyon with a cliff higher than Mount Everest.

Sometimes called the "Humpty Dumpty" moon, Miranda has uneven landforms that make it look like it was broken apart.

NASA/JPL-CALTECH

The Inverness corona, a chevron-shaped feature formed by an upwelling of ice, is its most distinguishing trait. Other well-known coronas are Elsinore and Arden. With gravity less than 1 percent of what we're used to on Earth, you can jump one hundred times as high. Daredevils might consider bungee jumping off of Verona Rupes, a twelve-mile-high cliff—the tallest in the solar system!

I'd rather be sailing

THE WINDS OF
NEPTUNE

Destination ♆ Neptune

Neptune is a planet of endless blue. Almost 3 billion miles from the sun, this ice giant appeals to those drawn by intense solitude and peaceful darkness. It's a secluded locale where you won't be bothered by droves of tourists. This radiant blue sea of churning gas punctuates the blackness. You'll be spellbound by Neptune's dramatic cloudscapes. It's slightly smaller, denser, windier, and bluer than Uranus, with a gravity similar to Saturn's. It has a strange magnetic field, fourteen moons, and a few scant rings. As you sail through its hazy atmosphere, or admire the vast blue from a nearby moon, you'll wonder why you didn't make the trip to one of the most majestic and mysterious planets in our solar system sooner.

Even though this hovering orb of blue serenity looks harmless enough, don't let the color hypnotize you into a stupor. Like Jupiter, Saturn, and Uranus, it is a planet of treacherous storms, and it's home to the fastest gusts in the solar system. You could spend days, weeks, or months sailing its exhilarating winds.

AT A GLANCE

DIAMETER: 3.88 times Earth's

MASS: 17 times Earth's

COLOR: Bluer even than Uranus

SPEED AROUND THE SUN: 12,000 miles per hour

GRAVITATIONAL PULL: A 150-pound person weighs
169 pounds

AIR QUALITY: Thick, 80 percent hydrogen, 19 percent
helium, 1.5 percent methane, trace water and ammonia

MADE OUT OF: Gas

RINGS: Yes

MOONS: 14

TEMPERATURE AT 1 BAR: -330 degrees Fahrenheit

DAY LENGTH: 16 hours and 6 minutes

YEAR LENGTH: 163.8 Earth years

AVERAGE DISTANCE FROM THE SUN: 2.79 billion miles

DISTANCE FROM EARTH: 2.67 billion to 2.92 billion miles

TRAVEL TIME: 8.7 Earth years for flyby

TEXT MESSAGE TO EARTH: 241 to 258 minutes

SEASONS: Long

WEATHER: Cloudy

SUNSHINE: 900 times dimmer than on Earth

UNIQUE FEATURES: Last planet in the solar system

GOOD FOR: Feeling blue

Weather and Climate

Neptune is as cold and windy as you'd expect the last planet in our solar system to be. At almost 3 billion miles from the sun, a temperature of -330 degrees Fahrenheit is fairly normal. This ice giant is a whirl of gaseous hydrogen, helium, and methane. It's the methane that gives it its beautiful blue complexion and one reason it's just a little warmer and a little bluer than Uranus. In spite of the cold, Neptune has a hot heart. The contrast between the superheated center and frigid exterior drives Neptune's forceful winds and storms.

On an average day, you'll have to navigate thick, slow-moving hazes made of ethane, hydrogen cyanide (which is known for its almond aroma), and methane high in Neptune's sky. As you descend, temperatures plummet and gaseous haze gives way to icy ammonia and water clouds. These clouds sometimes form bands that cast shadows on the blue abyss below.

As with other gas giants, the height where the pressure is the same as Earth's at sea level is a convenient reference point to orient yourself in the unsettling groundless sky that is Neptune. As you descend below that point, the hazes and clouds transition to an icy "ocean," though it's nothing like the oceans of Earth. Instead of a liquid layer, this ocean is a concoction of unbonded oxygen, nitrogen, carbon, hydrogen, and compounds of ammonia and methane that exist in a muddled space between solid, gas, and liquid. Pressures skyrocket to a hundred thousand times that on Earth's surface if you go deep enough. Descend far enough into Neptune's hot mantle and you'll eventually reach its small core, a hard ball of rock and ice.

The seasons here are longer than on any other planet that orbits our sun. Neptunian years last almost 165 Earth years, so a single spring lasts half a human lifetime. Neptune has a 28-degree tilt, and seasonal changes are noticeable in the way clouds brighten in areas that get more sun. It's windy all year, and the winds here approach supersonic speeds.

The change in the length of day as you move north or south, characteristic of all gaseous planets, is very noticeable on Neptune. Though the official length of day is 16.11 hours, that's actually just the average time it takes all the gas at different latitudes to rotate once around. The poles can zip around in just twelve hours, while it takes material at the equator almost eighteen hours. A thick central band of gas hugs the equator from 50 degrees south to 45 degrees north, moving eastward at speeds near 900 miles per hour. This is where you'll want to hang out if you want to stretch out the day.

Within Neptune's powerful yet steady winds, storms can appear suddenly as dark spots, marring Neptune's tranquil gases. They're immense vortices, bordered by radiant clouds. Unlike the storms of Jupiter, these storms form and fritter away more frequently.

When to Go

Considering its remoteness, the extreme cold, and the windy weather, there is no good time to visit Neptune. There is no bad time, either. For example, if you want to go during the northern hemisphere's summer, the season lasts around forty years. Even at the height of summer you'll never escape the life-threateningly frigid temperatures.

Though the particular season you visit doesn't matter much, storm watchers like Neptune when seasons are changing and there's a higher chance that storms will form. Neptune's next northern spring equinox is happening in 2044.

Getting There

As the last official planet in the solar system (sorry, Pluto!), Neptune is one of the most expensive planets to visit. It's far away and trudges around the sun at just 12,000 miles per hour (Earth moves at 66,616 miles per hour), so you'll have to pack a lot of fuel if you're interested in slowing down to get a closer glimpse.

To save fuel on the 2.7-billion-mile journey, you can get a gravity assist from Jupiter. The most dramatic and memorable part of your journey are these slingshot maneuvers, when you are flung by the gravity of a planet as it moves past you. Over the course of

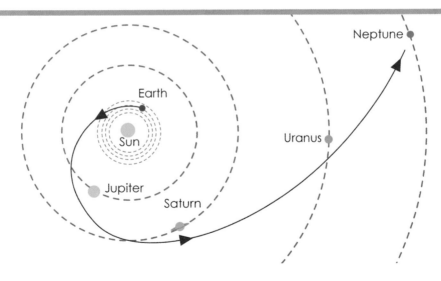

Getting to the final planet in our solar system will take some time, but the destination is worth the wait.

several months you'll be treated to gorgeous views of the gravitational assistant, Jupiter.

You can get a gravity assist from other planets too—there are only about two dozen times between 2020 and 2070 when they'll be available for a boost. Don't wait too long to book your flight. The average flight time is almost a decade, so the vacation package comes with a twenty-year minimum commitment. Best to start planning now, although with some newer—and highly experimental—ion drive technology, you may be able to get there in as little as two years.

When You Arrive

As you get closer to Neptune, you'll see that it's bustling with wine-dark storms that disrupt its pure blue glow. The first thing you'll do is pop down from orbit to see them up close. Within Neptune's atmosphere you'll be slightly heavier than you are on Earth, but by so little you probably won't even notice the difference after your long trip in microgravity. Some people love slipping back into the Earthlike gravity, while others find it oppressive. Whether you get to know the skies of Neptune in an airplane or an airship, steady yourself for some serious turbulence. Neptune's winds are the strongest in the solar system.

You'll soon become intimate with methane. It's what gives the planet its deep blue color, thanks to the way it scatters light. 1.5 percent of Neptune's atmosphere is made of this hazardous gas. If too much methane leaks into your suit or habitat it can start a dangerous fire, but it won't burn outside because Neptune's atmosphere has too little oxygen to fuel the flame.

Getting Around

You'll need an airship sturdy enough to bear the rush of Neptune's record-breakingly fast winds. Near the equator, they can top 1,200 miles per hour. Once you're at high speed, you won't notice how fast you're going. It's the moments when the wind changes that you'll realize you're in motion. At high altitudes, the breezes are slower. They increase as you descend or travel toward the poles. Like Uranus, Neptune has thick jet streams. The one that straddles the equator travels westward.

Neptune has the highest percentage helium of any planet in the solar system, 19 percent, and most of the rest of its atmosphere is the even lighter gas hydrogen. This, combined, with Neptune's gravity, means the atmosphere isn't very dense, so it's hard to keep an airship floating. A giant blimp needs to be filled with warm air or completely empty—a vacuum airship.

As an alternative, you can travel by airplane. As long as you have an engine that works without oxygen, you should be able to stay airborne.

If you plan to travel to Neptune's murky depths, you will need a robust ship rated for high pressures. It's best not to venture below the point where the pressure exceeds the equivalent of a 3,300-foot dive beneath Earth's ocean. Dive any deeper and your vessel will be unstable. Dealing with an imploded ship is unpleasant on a relaxing holiday.

Once you're ready to head to Neptune's rings, arcs, and moons, you'll need a rocket ship two times as powerful as the one you used to launch from Earth. The nearest moon, Naiad, is only

14,600 miles away. Triton, Neptune's largest moon, is about 200,000 miles, just a little closer to Neptune than the Moon is to Earth.

What to See

The Great Dark Spot

Jupiter's Great Red Spot gets all the attention, but Neptune has plenty of spots of its own. The Great Dark Spot of 1989 rattled the skies of Neptune, ballooning to roughly the size of Earth—the same

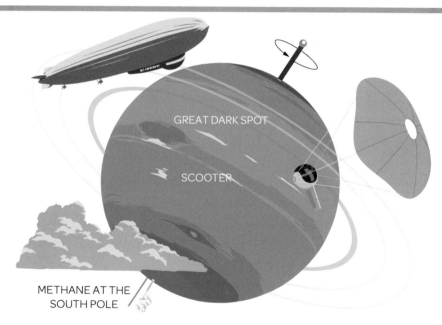

GREAT DARK SPOT

SCOOTER

METHANE AT THE
SOUTH POLE

Explore Neptune's azure clouds, storms, and spots.

relative size to Neptune as the Great Red Spot is to Jupiter. The eight-thousand-mile-wide dark blue void in Neptune's methane layer opened a window into the deeper, darker layers of the planet. Rocking the planet's atmosphere for several years, unlike Jupiter's Great Red Spot, it morphed and stretched in a predictable pattern, shrinking and lengthening over a cycle of eight days, like a breathing giant. Nowhere else has such a phenomenon been found. Spinning in the opposite direction of Neptune's rotation, the atmospheric vortex completed its journey around the planet every eighteen hours, drifting westward at 67 miles per hour. Jupiter's Great Red Spot moves at just 6 miles per hour relative to the air around it. Though smaller in width than the Great Red Spot, this pattern may have been deeper than the Great Red Spot is tall, according to storm watchers. Along the edge of it was a long cirrus cloud, its faithful companion.

The Northern Great Dark Spot

Though the Great Dark Spot is long gone, if you're lucky you'll be able to catch another one of Neptune's great spots. They come and go more frequently than Jupiter's spots, and stick around for a few years at a time. Still a remarkable fact, considering the longest hurricane on Earth lasted only thirty days.

The Northern Great Dark Spot, though slightly smaller than the Great Dark Spot, covered an area the size of New Mexico. It had small bright companion clouds bordering it. Since early robotic explorers returned images of Neptune in the 1980s, about a dozen great dark spots have been tracked. A large one the size of the United States was recently reported. You might also look for scooters, small white clouds that form and zoom through Neptune's upper atmosphere.

It's impossible not to reach for your camera and start snapping away at Neptune's spots. Just remember you'll have to wait eight hours to see if anyone on Earth "Likes" your picture of one of its storms.

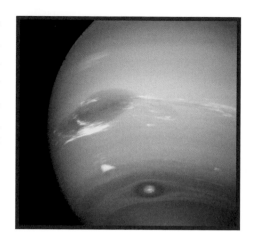

Neptune's storms may look benign from a distance, but they are some of the fiercest in the solar system.

VOYAGER 2 TEAM/NASA

Warm South Pole

Neptune is leaking methane from its south pole as if it ate too many beans. It's ten degrees warmer there than in other places, which lets the ice crystals suspended in the atmosphere turn to gas and escape. This behavior is seasonal. In about eighty years, the leaking methane will occur at the north pole, which will then be facing the sun during the northern summer.

Mysterious Interior

Since it's hard to survive the extreme pressures of Neptune's interior, there are many rumors about what lies there. Under the gaseous atmosphere, Neptune is more than an ice giant—it is a fluid giant. True to its namesake, the roman god of the sea, there is thought to be a pressurized, liquid water and ammonia ocean layer.

Activities

View Neptune's Rings and Arcs

You'll have fun speculating about what the early astronomers after whom Neptune's rings are named—Galle, Le Verrier, Lassell, Arago, and Adams—would think about humans basking in Neptune's blue glow while inspecting its arcs and rings up close. The rings are made of dark, reddish particles containing water ice and lots of dust. They're kept in place by Neptune's small inner shepherd moons. An additional unnamed ring of dust encircles the planet at the equator. Circling the planet alongside its sparse rings, Neptune's arcs, or not-quite-complete rings, have large gaps caused by tiny gravitational tugs from nearby moons.

Tour the Haze and Clouds

When you look at Neptune, you look through the clear atmosphere of its stratosphere into the troposphere, full of clouds and thick, slow-moving haze. The methane haze that hangs in Neptune's air is perfectly natural, though no less unpleasant than thick smog in Earth's crowded cities.

Cloud admirers will delight in comparing the formations on Neptune with those that float in Earth's sky. While they may seem like the familiar wispy, high-altitude cirrus clouds seen on Earth, these ones are made of swarms of methane ice crystals traveling at dangerous speeds. Lenticular clouds, shaped like lenses, form stratified layers that make for spectacular photos.

You'll recognize methane clouds by their blue tint. A thin, patchy layer of them resides just below the one bar level—the height where the pressure equals the pressure of Earth's air at sea level. Below that are condensed hydrogen sulfide and ammonia clouds. At around fifty bar (equivalent to a sixteen-hundred-foot dive into Earth's ocean) are clouds made of water.

Search Out the Elusive Auroras

Feel free to pack your compass, but don't expect to navigate with it. Neptune's magnetic field, though strong, is nothing like Earth's. It's close to the surface, and more complex than what we're used to. It is thought to be influenced by metallic hydrogen deep in the planet's interior. Conditions here are similar to those of the other giant planets. At these pressures, hydrogen can exist only in a metallic state rather than a molecular state. Electrons flow freely, and can mess with navigation.

The magnetic field is not lined up with the planet's geographic poles. Instead it's tilted 47 degrees from the rotational axis, pointing toward the sun. It would be like having Earth's North Pole as far south as New York City, Rome, or Beijing. Neptune's field is also off center, rather than in the middle like Earth's. The magnetic field is rather unpredictable. If you do bring your compass, it will drift haphazardly in different directions, making you feel like you're in an alien Bermuda Triangle (but this one is real).

All of this magnetic weirdness makes finding Neptune's elusive dim pink auroras all the more thrilling. As the magnetic field twists and turns, so do the auroras. They wander around the planet in unpredictable paths, so it's hard to know where to look.

Tune In to Lightning

Bring a portable radio and you can tune in to Neptune itself. It emits radio waves in the very low frequency range, between six and twelve kilohertz. The strength of the signal changes with the planet's rotation. Some times of day are better to tune in than others.

Lightning on Neptune is similar in power to Earth's. When it strikes, you may be able to use a very low-frequency radio receiver to catch a whistler. They're strong, high-pitched bursts that gradually lower after a lightning strike. Lightning strikes a hundred times an hour, so with practice you should be able to find one.

What's Nearby

Neptune has fourteen moons. Most are rocky fragments, the biggest not more than a few hundred miles across. About half are barely moons, really. Called captured objects or collision fragments, they're only a few miles wide and orbit millions of miles from Neptune, taking years to go around it just once. You can ignore those faraway specks of rock. Nobody really goes there.

Stick close to Neptune and explore its so-called regular moons, the inner moons, just tens of thousands of miles away with plenty of terrain to explore. From the vantage point of its inner moons, you get great views of the planet's faint rings.

Triton

Neptune's largest moon, Triton, is a winter wonderland covered in nitrogen, methane, and carbon monoxide ices, with a light dusting of

199

clean, fresh nitrogen and methane snow. As the ice has settled over time its craters and canyons have flattened out, meaning you won't have to worry about any steep hills, mountains, or deep gorges.

Triton is the seventh-largest moon in the solar system, at just sixteen hundred miles across, and it's smaller than Jupiter's Galilean moons, Saturn's moon Titan, and our own moon. However, it is the only large moon in the solar system to orbit its mother planet in the opposite direction of the planet's rotation, indicating that it's a celestial invader rather than native to the region. While most modern moons form at the same time as the planets they circle, moons like Triton are thought to be captured objects, wandering into the planetary neighborhood by happenstance. Triton was probably pulled in by Neptune's gravity as it was passing by on its way somewhere else.

Over time, Triton itself will sweep up Neptune's tiny inner moons. And in about 3.5 billion years, the whole moon will be torn apart by Neptune's gravity, making bright rings that may outshine Saturn's famous rings. Something to look forward to if you stay long enough!

If you visit Triton now, you'll find it's brighter, more reflective, and colder than most moons, with forecasts often predicting temperatures of hundreds of degrees below zero. Winds aren't too bad. They tend to flow westward at 11 miles per hour, so no need to worry about windchill. There are usually no clouds but plenty of haze. Small ice chunks crunch under your feet, covering a thick layer of dirty water ice. The atmosphere is tenuous. What amount of it that exists is 99 percent nitrogen, with traces of methane and carbon monoxide. The part of the atmosphere called the thermosphere gives off eerie light called airglow. It's caused by UV light interacting with charged particles in Neptune's magnetosphere,

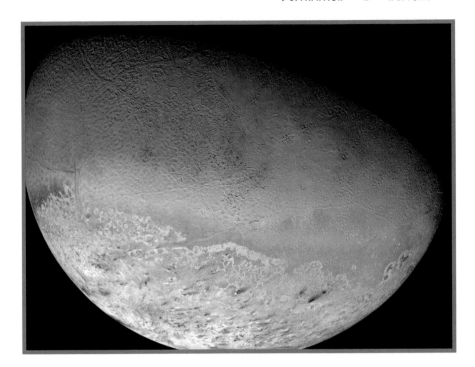

Cantaloupe terrain on Triton

NASA/JPL/USGS

which in turn interact with Triton's upper atmosphere.

Though the atmosphere is sparse, it's enough to make shooting stars visible. Beautiful bright meteors can be seen from the surface, and if you're lucky you may see one travel all the way to the ground.

The moon does have an atmosphere and tilt, and therefore, seasons and weather. But they are different from its mother planet's because the axial tilt is different. Seasons are long on Triton, and markedly different from the ones on Neptune. They are at times more extreme and less extreme than Neptune's. On Triton, hot and cold are in the eye of the beholder. A very hot summer on Triton may have an average temperature of -388 degrees Fahrenheit.

Missing fresh fruit on your long journey? Visit the Bubembe region, where you'll get the best view of the moon's cantaloupe terrain. You can tour it along twisting sulci, which may remind you of the tiger stripes of Enceladus. The ten-mile-wide grooves continue for hundreds of miles. They have gentle slopes, similar to that of a wheelchair ramp. The best ones to visit are Boynne Sulci in the south and Slidr Sulci in the north.

East of Bubembe is the Monad region, with rugged fossae, or trenches, that might look familiar if you've traveled to Mercury. Look for the strange mushroom features named Zin and Akupara.

The plains of Ruach and Tuonela are surrounded by walls hundreds of feet high. They are completely smooth except for pits and a crater or two. If you walk these plains, wear your Wellingtons; slushy ice may have made its way to the surface.

Uhlanga, near the southern pole, is covered by a pink cap. A darker, redder edge is probably due to UV light interacting with methane. Tiny crystals of methane ice scatter the incoming sunlight. Here, it's eternal midsummer. The region is in sunlight for more than a hundred years, though it looks darker than Earth's night.

One of Triton's greatest attractions is its nitrogen ice geysers. A layer of liquid nitrogen a hundred feet below the surface feeds them. Though the moon is freezing, its pressure is high enough to melt nitrogen. When the pressure falls to a tenth of its normal atmosphere, the nitrogen rises through a geyser nozzle at more than 300 miles per hour, sending material up for miles. You'll see the geysers Hili, named for a Zulu water sprite, and Mahilani, named for a Tonga sea spirit, 50 degrees south of the equator, clustered together with at least two other active geysers. More than a hundred dark spots dot the southern hemisphere. These are geyser remnants, ranging from a few miles to tens of miles wide.

Nereid

Nereid is Neptune's third-largest moon. It has a strange orbit. The eccentric shape of its path around Neptune brings it as close as 850,000 miles to the planet and as far away as almost 6 million miles. Spherical in shape, it's an icy moon and circles Neptune every 360 days.

Proteus

An early impact between this moon and another body nearly destroyed it, leaving behind a 150-mile-wide basin. Near the basin is another crater, fifty miles wide. Since this satellite is not round, it's more a collision fragment than a standard moon. There is another depression in the southern hemisphere called Pharos. It's 160 miles across, with raised rims and a flat floor six miles from the rim. The moon itself is just over 250 miles wide, meaning whatever left this depression might have nearly destroyed the weirdly shaped moon.

Destination ♇ Pluto

Pluto, the celestial body formerly known as a planet, is the most beloved—and controversial—space vacation destination in the solar system. Though it has lost its prominence in the pecking order of space objects, Pluto is still—and always will be—the same celebrated getaway spot favored by those who crave the most isolated locales. Discovered in 1930 by astronomer Clyde Tombaugh, this hunk of ice has captured the imaginations of generations. Named for the Roman god of the underworld, if *Pluto* is another word for *hell*, it seems to have frozen over.

If you're like most ordinary people, you've always dreamed of visiting Pluto, but like a lot of them, you've never gone. Located in the distant and ghostly Kuiper Belt, at times almost 5 billion miles from Earth, this land of rock and ice, smaller than Earth's moon, marks the beginning of the end of our solar system. You'll be captivated by its rugged pink-hued mountains, deep dark blue sky, and iconic Tombaugh region, a huge icy plain in the charming shape of a heart. Water ice and nitrogen glaciers slowly drift across Pluto's surface over century-long seasons, which are notable despite the dwarf planet's tremendous separation from the sun. Its cratered, pitted terrain and mile-high mountains provide ample amusement for explorers. The gravity is low—you'll weigh less than half what you do on Earth's moon—and you can glide like a feather across its frigid fields.

P AT A GLANCE

DIAMETER: .2 Earth's

MASS: .2 percent of Earth's

COLOR: Peach, gray, deep rust

SPEED AROUND THE SUN: 10,500 miles per hour

GRAVITATIONAL PULL: A 150-pound person weighs
9.5 pounds

AIR QUALITY: Barely there with traces of nitrogen, methane,
and carbon monoxide

MADE OUT OF: 70 percent rock, 30 percent ice

RINGS: None

MOONS: 5

AVERAGE TEMPERATURE: -369 degrees Fahrenheit

DAY LENGTH: 153 hours

YEAR LENGTH: 248 Earth years

AVERAGE DISTANCE FROM THE SUN: 3.67 billion miles

DISTANCE FROM EARTH: 2.66 billion to 4.68 billion miles

TRAVEL TIME: 9.5 Earth years for flyby

TEXT MESSAGE TO EARTH: 238 to 418 minutes

SEASONS: Long and intense

WEATHER: Cold

SUNSHINE: Very dim, from 0.04 to 0.1 percent of Earth's

UNIQUE FEATURES: The heart-shaped Tombaugh region

GOOD FOR: Supreme solitude, a truly deep freeze

Weather and Climate

Pluto is cold, even by outer solar system standards. Winter, spring, summer, or fall, the temperature hovers between a soul-freezing -360 and -400 degrees Fahrenheit. Gases rising from the nitrogen ice and water ice that cover the surface make it even colder, like sweat cooling your skin. Unless your space suit is well insulated, anything you touch will instantly turn from ice to gas. Instant frostbite, especially from your toes losing heat to the ground, is a looming threat.

The weather on Pluto doesn't change much from day to six-Earth-day-long day because of its paltry air, made from nitrogen, methane, and carbon monoxide emanating from its icy surface. The air pressure is one hundred thousand times lower than on Earth, and if you squint your eyes you can see the atmosphere in the form of faint bright lines stretching across the dark sky. You won't get caught in a blizzard or feel gusts of wind, but you might see a meager low cloud. Pluto circles the sun just once every 248 Earth years, so one season lasts most of a human lifetime. Its orbit is a stretched ellipse, and at its farthest point it is almost twice as far from the sun as it is at its closest point. It's flipped almost upside down compared to most proper planets, with a tilt of 120 degrees. This sharp tilt causes land near the outward-facing pole to be cast in darkness for hundreds of Earth years at a time. When Pluto is far from the sun and where the surface is in shadow, you'll notice epic frosts covering the ground as much of Pluto's thin air freezes solid. These are the longest, coldest, darkest winters in the solar system. At the other pole, you are sunlit for centuries, although you definitely won't be sunbathing.

Because of the cold, most people don't realize that Pluto has its own form of the greenhouse effect due to the bit of methane in its air. Though the temperature is always uncomfortably chilly, relatively small differences can seem dramatic to those familiar with the Plutonian landscape.

When to Go

Since Pluto has such a wide-ranging orbit, it's best to try to catch it when it's closest to the sun. Earth's orbit is practically right next to the sun from Pluto's perspective, so this is when Pluto is closest to Earth as well. If you didn't make the trip the last time this happened, in 1989, you'll have to wait until Pluto's next approach, in 2237. You can buy a ticket for your great-great-great-great-great-great-grandchild! Even the quickest route will take a decade of your life—at least two if you're hoping to make it a round trip. Schedule your vacation when you are young, so you can return to Earth by middle age. Another option is to wait for retirement. Live out your last days wandering through the dwarf planet's cold plains, far away from earthly woes. No matter when you choose to leave, remember to pack your warmest clothes, along with your space suit. It is always far colder on Pluto than the deepest arctic winter on Earth.

Getting There

Getting to Pluto is a lot faster if you don't stop once you arrive. A direct flyby with chemical rockets takes anywhere from eight to twenty years, depending on your trajectory and how far away Pluto is from Earth. You can snap a few photos as you speed past on your way to the great icy space rocks that lie beyond the dwarf planet. If a mere flyby doesn't satisfy you, your trip planning gets more complicated.

Pluto moves much slower around the sun than Earth, at around 10,446 miles per hour, and slowing down to match pace with it takes a lot of energy.

A Hohmann transfer orbit works well for the inner planets with shorter orbits around the sun. Out here, that simple elliptical path would take many decades to complete because of Pluto's long year. A gravity assist from Jupiter can shave a few years off your trip, but let's face it—the trip to Pluto will be a long one unless you opt for a nuclear-powered rocket. Forget solar-powered travel. It doesn't work well this far from the sun. Because of the long travel time, you may opt to settle permanently.

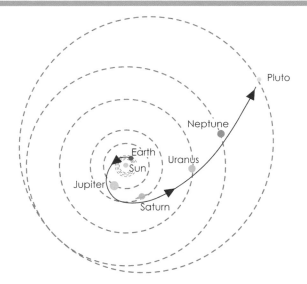

Though Pluto is far away, there are plenty of planets along the way to give visitors a gravity boost.

A view of Sputnik plains, the bright, heart-shaped area that is the most famous landmark on Pluto

NASA/JOHNS HOPKINS U. APPLIED PHYSICS LABORATORY/SWRI

When You Arrive

As your view of Pluto grows, you'll start to see the contrast between the dark stains and bright splotches on the surface, including Pluto's famous heart. You'll also make out Pluto's moon Charon and the deep chasm that slashes through its face.

Arrival is a big event when you've been restricted to a tiny capsule for years on end. Some are hesitant to leave the cozy accommodations of their spaceship. When you finally venture out you'll be greeted by a wintry panorama.

The cold, virtually airless environment makes for crystal clear skies, which are dusted with stars both day and night. The sun is always the brightest star in the sky, hundreds of times brighter than the full moon on Earth. Throughout Pluto's year as it draws closer to the sun, the sun grows larger in the sky. It appears four times as bright at its closest than at its farthest point. On the horizon, in the direction of the sun, the skies are a deep blue that fade to black as you look up. Sunsets on Pluto are otherworldly. They last for hours, since Pluto's day is six times longer than Earth's. Just after the sun sinks below the horizon it casts blue beams in every direction.

Don't forget to send a note to friends and family back home when you arrive. You'll need some extra patience when sending messages from Pluto, and because Pluto's orbit is so elliptical the time to send a message varies a lot more than it does for the planets. Depending on where Pluto is when you arrive, it will take between three and seven hours for your note to make it back. If you're feeling homesick, point a powerful telescope back at Earth and you'll see what happened a few hours ago because of the time it takes for light to travel. It's a telescope into the past.

Getting Around

Long-distance travel over the Plutonian surface requires rugged rovers than can manage jagged ice. Hoppers are useful because they can avoid this treacherous terrain. In low gravity, they'll cover long distances, and the thin atmosphere doesn't offer resistance. If you want an aerial tour of Pluto, you won't be gliding in an airplane; you'll fly the skies in a rocket.

A form of travel unique to icy worlds such as Pluto is the hover car. By allowing ambient heat from your comfortable, pressurized vehicle to escape through the bottom of its frame and heat the icy surface, the hover car can create a cushion of air. You'll glide over the smooth plains with very low friction.

Jaunting between Pluto and its small moons is easy. Because of its low gravity and lack of atmosphere, you don't need much fuel to escape its pull. You only need to reach a speed of 2,700 miles per hour to launch back into space. Traveling to nearby Charon is like traveling halfway around Earth. It's twelve thousand miles away, so you could be there in less than an hour if you launch at conventional Earth launch speeds.

What to See

Tombaugh Region (Tombaugh Regio)

Visitors to Pluto are eager to view the bright plains of the famous heart-shaped Tombaugh region. The western lobe, an icy depression known as Sputnik plains (Sputnik Planum), is miles deep and five hundred miles wide. Huge cracks separate the plain into geometric shapes, like the cracks in the sand of a parched desert—be careful not to fall in. These formations are thought to be the tops of areas of slowly churning nitrogen ice—like a superslow and very cold lava lamp. Because water ice is less dense than nitrogen ice, occasionally you'll see a huge chunk of water ice that has made its way to the surface and seems to float like an iceberg in the solid nitrogen. As you navigate this area of Tombaugh, you'll see dozens of distinct landforms, from rugged pits to bright plains and isolated hills.

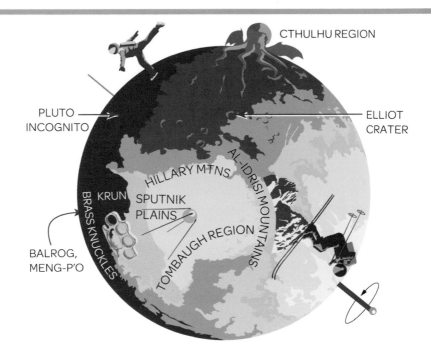

CTHULHU REGION

PLUTO
INCOGNITO

ELLIOT
CRATER

HILLARY MTNS

AL-IDRISI MOUNTAINS

KRUN SPUTNIK
PLAINS

BRASS KNUCKLES

TOMBAUGH REGION

BALROG,
MENG-P'O

Your favorite icy dwarf planet has plenty of wintry fun to explore in every season.

Al-Idrisi Mountain Range

Over the western curves of the heart, a chaotic mountain range looms. Ice climbers will appreciate these mountains, which tower as high as the Rockies. The water ice mountains have a delightful pinkish hue, with white snowcaps. The pink is methane that has worked its way into the pristine white, like dirty snow. The snow is unfortunately ill-suited to melt for drinking water, since it's made out of frozen methane. On Pluto, the saying goes, "If the snow is pink,

213

don't drink." Flowing between the peaks are glaciers of frozen nitrogen, which in gas form makes up most of Earth's atmosphere. Take a good look—this is what frozen air looks like.

Hillary and Norgay Mountains (Hillary Montes, Norgay Montes)

Travel southwest from Sputnik plains to visit the Hillary Mountains and Norgay Mountains. They are named after Edmund Hillary and Tenzing Norgay, the first recorded climbers to summit Mount Everest. These mountains aren't nearly as tall as the Himalayas, where Everest stands. Norgay's tallest mountain is just 11,000 feet from its base, compared to the summit Norgay and Hillary achieved on Everest: 29,029 feet, or 15,000 feet from base to peak. Don't be discouraged if this feels less impressive than the accomplishment of

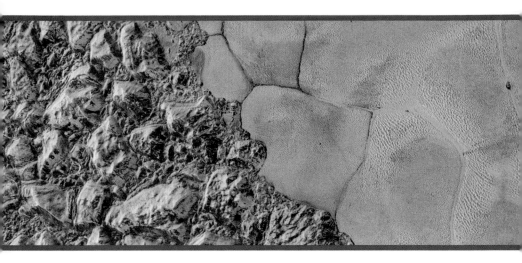

On the left are the pink-hued water ice mountains of the al-Idrisi range, while the right shows the edge of Sputnik plains.

NASA/JOHNS HOPKINS U. APPLIED PHYSICS LABORATORY/SWRI

those British mountaineers in 1953. Remember, you already traveled at least 2.7 billion miles to get here.

Brass Knuckles

Between the Cthulhu region and Pluto's heart, Pluto has vast dark regions lined up like the finger holes on brass knuckles. Individual spots are named after mythological creatures, and include Meng-p'o, the Chinese goddess of forgetfulness, Balrog, the subterranean demon of *The Lord of the Rings* fame, and Krun, the louselike lord of the underworld in the Mandaean tradition.

Cthulhu Region (Cthulhu Regio)

Named after the gigantic octopus-like monster from H. P. Lovecraft's stories, Cthulhu is actually shaped more like an enormous whale.

When you visit, you'll be hiking on very dark, icy ground that's blanketed in a tar-colored layer of carbon-rich muck called tholins. These highlands of Pluto are much older than the icy plains of Sputnik, and have had more time to accumulate craters—the scars of time. You'll enjoy exploring their mysteries. The fifty-six-mile-wide Elliot crater has a strange, bright ring of ice. Standing on the ring, you'll look up at its two-mile-high central peak, towering like a castle in the middle of an icy moat.

Activities

Ski

If you're looking for something a little more thrilling than your typical resort vacation, try the backcountry descents on Pluto, which rival any black diamond routes back home. Heated skis create a layer of vapor, which reduces friction with the snow. Skiing is easiest on the methane snowcaps, where you're more likely to find Pluto's version of powder. At only one-sixth the mass of Earth's moon, Pluto's low surface gravity means a typical jump will launch you more than twenty-four feet into the air, and you'll come down with a soft landing. There are plenty of steep jumps to vault off. Speed demons will need to have patience; the low gravity makes it hard to accelerate quickly. If you ski downhill long enough you can eventually get up to breathtaking velocities, especially because there is no air resistance. You can even catch some air, though technically there is no air to catch.

Ice-skate

Ever try ice-skating on rock? That's what frozen water ice is like on Pluto, where it exists in an ultradeep freeze at about −400 degrees Fahrenheit. You'll have better luck skating on nitrogen ice rinks, because nitrogen has a much lower freezing point than water: -346 degrees, much closer to Pluto's ambient temperatures. This new form of ice-skating may take some getting used to. When you ice-skate on Earth, your skates glide over a layer of liquid. On Pluto, you might want to try heated blades, which will turn the nitrogen ice into gas to give you a smooth skate.

Mountain-Climb

The pink mountains of Pluto have cliffs that may seem rugged, but they are surprisingly easy to scale because of the low gravity. You'll need strong ice-climbing gear, including crampons and ice picks, to tackle them. People are sometimes cavalier about climbing safety in low gravity, but don't forget that although you can fall from higher heights without hurting yourself, you can still die from a long enough fall.

Cliff Jump

On Pluto, you can jump from a hundred-foot tower without breaking your legs, due to the dwarf planet's weak gravity. Just after you jump you'll float downward gently, like a snowflake. Though it may seem like you could fall like this forever, you will accelerate as you near the ground. By the time you reach it, you'll only be going as fast as you'd be going if you'd jumped from a six-foot wall on Earth. You need no parachute—and even if you had one, it wouldn't do you any good because Pluto has almost no atmosphere. A popular location for cliff jumping is in the wide, deep pits that dot the Sputnik plains.

Play Air Hockey

If you heat a puck to Earth's room temperature on Pluto, it will instantly and violently turn the ice beneath it to gas; the surface of the dwarf planet becomes your personal air hockey rink. Just be sure to keep the puck moving or it will make a hole in the ice and disappear.

Debate Pluto's Dwarf Planet Status

There is nowhere more fitting to get into an impassioned debate about Pluto than on the dwarf planet itself. What's in a name? According to the International Astronomical Union, a planet is a celestial body that satisfies three criteria. First, it must orbit the sun. Second, it must have enough mass that it becomes nearly round under its own weight. Third, it should have cleared the neighborhood around its orbit, meaning there aren't other objects that share the same path around the sun.

This last criterion is where Pluto fails; it shares its orbit with many other icy worlds. It is also an oddball in other ways, including the fact that its orbit is highly tilted compared to the flat plane the planets orbit in. In addition, there are many objects in the solar system that are similar to Pluto (Neptune's moon Triton is bigger than Pluto), but were never considered planets. The choice was to downgrade Pluto, leaving the solar system with eight planets, or keep Pluto as a planet and possibly add many more to the solar system's map.

Regardless of the rationale, Pluto's demotion to dwarf planet status has advantages for the savvy space traveler, including discounts, making it a money-saving and peaceful place to spend your vacation.

What's Nearby

The main moon of Pluto you may have seen in pictures is icy Charon. The dwarf planet also has a group of smaller moons named Nix, Hydra, Kerberos, and Styx. The football-shaped moons Nix and Hydra are confusing to visit because they have unpredictable rotation. If you stay awhile on these moons you'll find the day varies in length and there are dramatic changes in the direction from which the sun rises. Their slow, chaotic spins are due to their interaction with Pluto, Charon, and the other moons; the gravitational influence causes erratic wobbling.

As you venture farther from Pluto, you may choose to visit several other dwarf planets or some of the many small, icy comets distributed in the vast ring-shaped region around the sun called the Kuiper Belt. Pluto, its moons, and all of these distant icy places are known as Kuiper Belt Objects.

Charon

You'll see that Charon looms large in Pluto's sky, at twice the size the full moon appears on Earth. Each time Pluto rotates once, Charon orbits once. Half of Pluto's surface never sees the moon; likewise, half

of Charon never sees Pluto. While it's more expensive to stay on the side that faces the moon, the extra cost is worth the view.

The distance from Pluto to Charon is the same as from Shanghai to Buenos Aires—the space equivalent

Pluto's moon Charon

NASA/JOHNS HOPKINS UNIV. APPLIED
PHYSICS LABORATORY/SWRI

of a puddle jump. Pluto and Charon are more like two stars orbiting one another than a moon orbiting a (former) planet, because they have similar mass. They are trapped in a gravitational dance, like two people holding hands and spinning together in a circle. Because Earth is so much more massive than the Moon, it makes sense to think of the Moon as moving and Earth as standing still. In reality, Earth and the Moon are spinning about a point called their center of mass. In the case of Pluto and Charon, the center of mass is well outside Pluto, so both Pluto and Charon appear to move a lot as they orbit each other. Try stationing your ship between them so you can watch them both circle around you.

Visit the Serenity chasm (Serenity Chasma), a huge canyon that stretches around the moon. It is longer and deeper than the Grand Canyon. It probably formed when an ocean on Charon froze and expanded, creating deep cracks in the landscape. You'll also want to visit Charon's mountain in the moat. Set in the center of a deep pit, this unique mountain has an enigmatic origin.

As you travel to different regions of Charon, you'll notice that the color changes. Near the moon's equator the ground is brighter than the darker, redder poles. A particularly dark region near Charon's north pole has been nicknamed Mordor after the menacing region in J. R. R. Tolkien's *The Lord of the Rings*. One does not simply walk into Mordor, since it is difficult to walk in only 3 percent gravity.

Haumea

Pluto and Charon are not the farthest places you can go to get away from it all. Other dwarf planets haunt the remote reaches of the solar system. Haumea is farther on average from the sun than Pluto, and is one-third its size. Named after the Hawaiian goddess of childbirth, the dwarf planet resembles the elongated head of a newborn, measuring twice as long one way as the other. Its unusual shape is due to its superfast rotation—a day on Haumea is just four hours long. Standing on it, you can detect the daily motion of the distant sun and stars as they rise and set, like watching a time lapse of a night on Earth. Its two moons, Hi'iaka and Namaka, are named after Haumea's daughters, the patron goddess of Hawai'i and the goddess of the sea, respectively.

Makemake

A remote world that orbits between 38 and 53 times the average distance from Earth to the sun, Makemake makes Pluto seem positively cosmopolitan. It is named after the creator of humanity and fertility god of the people of the island of Rapa Nui. This island is known to some as Easter Island, and Makemake, which has a single moon, was first discovered shortly after Easter.

Eris

This dwarf planet was originally called Xena, and its moon Gabrielle after characters from a television series. It was soon renamed Eris after the Greek goddess of strife and discord, and its moon Dysnomia after the daemon of lawlessness. Back when it was discovered, Eris led to much controversy in the classification of planets. Pluto was still considered a planet at the time, and Eris was similar to Pluto in many ways. If Pluto was a planet, shouldn't Eris be as well? After all, it is slightly more massive than Pluto. Instead of expanding the planet club, astronomers kicked out Pluto, redefining Pluto, Eris, and any similar objects as dwarf planets.

For the would-be visitor, Eris's distance is what sets it apart. At its farthest, it orbits at ninety-seven times the average distance from the sun to Earth. If you miss its closest approach to the sun, you'll have to wait a whopping 557 years for its next pass. Before you embark for Eris, we recommend reflecting on the depth of your commitment to solitude.

67P/Churyumov-Gerasimenko

Dark and lumpy in shape with two main lobes, and about as long as Mount Fuji is high, this comet is named after its discoverers, Klim Ivanovich Churyumov and Svetlana Ivanovna Gerasimenko. Mercifully known as 67P for short, it is on a journey from its origin in the Kuiper Belt, the home of Pluto, to the inner solar system. The minimal gravity warrants extreme caution—be careful not to come in too hot on approach, or you might bounce right off its dark surface. Refrain from jumping during your visit or you might jump into oblivion, never to fall back down. 67P has the honor of being the first comet to host a robotic visitor, the Philae lander, as a part of the *Rosetta* mission.

Like all comets, when it approaches closer to the sun than the orbit of Saturn, some of 67P's ice begins to turn to gas. This creates a bright cloud around the otherwise dark icy nucleus, which reflects light and makes it hard to stay on the comet unless you are well secured. Standing on 67P as it approaches the sun, you'll notice two tails coming off of it, the gas tail and dust tail. The gas tail points

ESA/Rosetta

*67P, the first comet
with a robotic visitor*

ESA/ROSETTA/NAVCAM

in the direction away from the sun because it is pushed away by the solar wind, while the dust tail often curves and points somewhere between the path of orbit and the gas tail.

While 67P won't cross Earth's path, some other comets do. When this happens, they leave a trail of dust particles, like bread crumbs, behind them. Each little grain of dust makes a gleaming streak across the sky, together making a meteor shower. Make a wish!

HOMECOMING

All good things must come to an end. As the date of your return to Earth looms closer, your anticipation will build. Keeping busy can help with the wait. Reconnect with your friends and family back home, read guidebooks about Earth, and visit your favorite tourist spots one last time. Know that from the earliest era of space travel, astronauts have emphasized how their journeys led them to appreciate Earth even more.

As you begin your journey home, the pale blue dot that you've watched in the starry skies from alien spheres will grow larger. You'll feel the ache of an epic journey coming to an end, and the blissful realization that you'll soon rest your head in your own bed, in your home gravity, for the first time in perhaps years or even decades. As you near Earth, you'll see the razor-thin glow of the atmosphere lining the globe, and wonder at just how important this seemingly small feature is to the trillions of life-forms beneath it.

The transition back to being an Earthling isn't always smooth. British astronaut Tim Peake described the process of returning to Earth after a six-month trip as the "world's worst hangover." The longer you've been in space, the longer it will take you to adjust physically and mentally to life on Earth. You may be weak and uncoordinated from living in microgravity, and it can take years to recover your prior strength. Walk with care and avoid impacts after you return home.

Your weakened bones, particularly your hips, may break under stress. In the first days and weeks back, you might have difficulty walking or be a bit clumsy. You'll forget that you can't gently push off any surface to get yourself across the room. More than one wineglass has been broken when someone forgot that he couldn't release his grasp of it for a moment while fetching a fresh bottle of wine. Things must be set down on surfaces here on Earth.

Many long-term space travelers experience culture shock. Depending on how long you've been away, the Earth you return to can be very different from the one you left, making you feel like a time-traveling outsider. Your taste in music and fashion might lag years behind the dominant culture's.

If you've been away for a very long time, you may have forgotten what it feels like to live in a completely natural environment that supports human life. Months, years, or even decades of living in climate-controlled habitats will leave your head spinning when you confront the fluttering weather outside on a typical autumn day, when temperatures can sometimes rise and fall ten, twenty, or even thirty degrees. Whether you're from the hottest deserts of the Sahara or the coldest corners of Siberia, you may even find yourself enjoying the extremes of your native environment for the first time, excited to strip down to your underwear to keep cool in the stifling heat or the chore of scraping ice off your car windshield at five o'clock in the morning.

It's possible you'll decide to relocate when you return, settling somewhere near the equator, where it's warm year round and you never need a coat. Or you might be excited to see how you fare at Earth's poles, where survival is tough by Earth standards but comparatively easy by alien standards.

Regardless of where you end up, you'll most likely enjoy Earth's four distinct seasons like you never have before, reveling in the dirty snow-covered streets in winter, the slushy mud of spring, the sticky days of summer, and the bitter wind gusts of fall.

No matter what time of year you return, or what type of weather you encounter, you'll be surprised at how excited you become over little things. You may find yourself leaving home without a jacket or even shoes in poor weather, because you want to feel everything that nature has to offer. You'll be amazed at friends who complain about the cold, or the heat, or the storms, which seem mild in comparison to the space weather you faced during your holiday away from Earth. You'll probably find it hard to stop looking at the big blue sky and white puffy water clouds within it. When the sun goes down, you'll stare at the blackness of the night and pinpoints of stars, and marvel at the places you have been. It will be challenging to keep yourself from getting into long conversations with strangers about the beauty of the universe, the fragility of humankind, and the urgency with which we need to do everything we can to protect our home planet.

On the other hand, you may become quickly irritated with nuisances of getting around on Earth. In cities, traffic jams are common, especially at certain times of day. Airports are often crowded, and you will be regularly subjected to security checks. Travel by water can be slightly more pleasant, but it is slow and service is less regular. Do your best to plan travel during off-hours, when fewer of the 7 billion people of Earth are on the go. Any kind of crowd may leave you breathless and prone to a panic attack, especially if you've been on vacation in a more remote area of the solar system where there were few people.

You'll be struck by Earth's stunning variety of natural beauty. You may have forgotten what the sound of rushing water was like, or that huge swaths of land could be covered entirely in lush greenery. Forests may make you anxious at first, and you may feel more comfortable in wide-open deserts. It will be hard not to compare craters, canyons, mountain ranges, and volcanoes to those you saw on your trip. Now that you're back, spend time enjoying Earth! Many natural features on our planet rival the sights you took in during your space vacation, though anyone returning from Jupiter will always long for its bright auroras, which dwarf the best of Earth's.

After every great vacation, returning to normal life can be bittersweet. More than likely, you'll have to change jobs because of extended absence. Take this time to reevaluate what is important. You may decide to commit yourself to improving conditions for the people of Earth, or to preserving the delicate environment that so effectively keeps the inhabitants of the planet alive. No matter what you decide to do with yourself as you get back into the routines of life on Earth, look back on your space vacation as the one time when you really, truly left it all behind. When life gets stressful, look up into the night sky, and picture yourself among the stars.

ACKNOWLEDGMENTS

This book would not exist without the geniuses behind Guerilla Science, an organization that pushes the boundaries of imagination and provides us with a nurturing outlet for our weird ideas. We are eternally grateful to Jen Wong, Mark Rosin, and Zoe Cormier for creating this unique collective in a muddy field in England all the way back in 2008. We are additionally indebted to Louis Buckley, Jenny Jopson, Sarah Barker, Kyle Marian Viterbo, Rachel Karpf Reidy, Pigalle Tavakkoli, and Marissa Chazan for helping keep the torch burning. A special shout-out is due Jenny Jopson, who came up with the original vision for the Intergalactic Travel Bureau, a place where the "retro futurism" of old-school sci-fi could flourish. Mark Rosin has been an especially dear adviser, and instrumental in getting the Intergalactic Travel Bureau off the ground in the United States.

Dozens of people have helped make our unassuming little space travel agency possible at festivals, museums, and disused storefronts in the United States and the United Kingdom. To all of our volunteers, astronomers, astrophysicists, and space travel agents, we are humbled by your tireless commitment to delighting thousands of unsuspecting passersby. We are deeply grateful to Steve Thomas, whose visionary art brings the Intergalactic Travel Bureau to life. Your talent is a constant source of inspiration and without your creations the Intergalactic Travel Bureau would be a cold corpse. We'd especially like to thank Steve for his inhuman patience as he forged art for this book in the face of our endless stream of tweaks, double reversals, scientific nitpicking, and failed creative experimentation.

Special recognition goes to Ferris Jabr, who was there from the beginning and always willing to lend a helping hand, both physically and spiritually, as an intergalactic travel agent and book adviser. To Colleen Cox, who helped lug an ungodly amount of gear onto a ferry boat for our U.S. debut at Governors Island in NYC. To astrophysicist Hanno Rein, one of our first U.S. agents and the creator of the amazing Exoplanet app, which gives visitors to the Intergalactic Travel Bureau an appreciation for the variety of fascinating exoplanetary worlds. To Renée Hložek and Lucianne Walkowicz, agents extraordinaire who provided invaluable support in our early days. To Janusz Jaworski and all the good folks at the chashama arts organization, who provided space for us to create. To Kaitlin Prest and Mitra Kaboli, who helped open up our first pop-up in Manhattan. To Zach Kopciak, the fantastic producer who brought flair and a level of unprecedented customer service to the experience. To our scientists Or Graur, Viviana Acquaviva, Steven Mohammed, Juan Camilo Ibañez-Mejia, Adam Brown, Lewis Dartnell, Adam Stevens, Ann Posada, Sarah Pearson, Federica Bianco, Kurt Hill, Robin Roberts, Andrea Derdzinski, Zephyr Penoyre, and all the other scientists, actors, and volunteers who joined us in San Francisco, Washington, D.C., New York, and London to plan space holidays. To the MetroPCS guy, who provided sound advice on how to get hardened New Yorkers to consider a vacation to the Moon (and/or switch their phone service). To

Linn Splane and Celena Tang, who brought style and grace to alien travel photography. To all of our Kickstarter supporters, clients, and the thousands of visitors who have sent postcards from space, taken pictures with us on the Moon and Mars, and pushed us to consider the infinite mysteries of vacationing in the great void.

We have been honored by all the people who devoted their time, energy, and support to launch this book. We are especially appreciative of the perpetually patient James Hedberg, whose shrewd critique and physics expertise guided our writing, and whose cooking nourished our bodies during the birth of this book. We are grateful to all the scientists and space industry experts who gave interviews or answered questions, including Richard Schmude, Ted Southern, Jonathan McDowell, Joan Hunter, Paul Spudis, Taka Tanaka, Gil Costin, Matt Heverly, Jim Papadopoulos, Geoffrey Landis, Keegan Kirkpatrick, Sarah Fagents, Robert Strom, Denton Ebel, Katherine de Kleer, David Blewett, Tom Stallard, Andrew Ingersoll, Mark Lemmon, Michael Person, Jani Radebaugh, P. J. Blount, Jessica Raddatz, Emily Rauscher, Paul Schenk, Michael Busch, and Tristan Guillot. Thanks to Amanda Moon, who provided valuable feedback and support as we considered how to turn an experience into a book. Thank you as well to all the talented writers and scientists of NeuWrite, who workshopped early chapter drafts.

We are indebted to the National Aeronautics and Space Administration (NASA) for the amazing data and high-resolution imagery it provides free for use in the public domain. Note that NASA does not endorse this book in any way, and neither do any of its affiliated institutions.

Special acknowledgment goes to Mara Grunbaum for her support and for introducing us to our agent, Rachel Vogel. Thanks to

ACKNOWLEDGMENTS

Katie Peek for introducing us to the art of Steve Thomas and giving design suggestions, and to Irene Pease (the Friendly Neighborhood Astronomer) for checking sections of the book for accuracy. We are indebted to our editor Meg Leder and the entire team at Penguin Random House, who have been nothing short of spectacular from the get-go in their enthusiasm for this project.

Jana Grcevich would like to thank her colleagues at the American Museum of Natural History, especially Mordecai-Mark Mac Low, Neil Tyson, Ashley Pagnotta, Statia Luszcz-Cook, Brian Abbott, Carter Emmart, and the library staff. She would also like to thank her friends and family: Geoff, Sarah, Mira, and Greg Grcevich; Stephanie Wykstra; Dawn Chan; Jocelyn Sessa; Josh Peek; Mark Wheeler; Beckie Wood; Aletta and Richard Tibbetts; and Kristin and Logan Lewis for their essential encouragement and companionship. She continually revels in how lucky she is to have such amazing people in her life.

Olivia Koski would like to thank James for cooking, cleaning, advising, and tending to her and to their newborn babe with little complaint during the extended periods when her face was fixated on a laptop screen. She additionally thanks her parents, siblings, in-laws, extended family, and many cherished friends for their support.

REFERENCES

In addition to conducting interviews with scientists and other experts to create this guide, we read dozens of books; combed through a trove of NASA technical reports, researcher blogs, scientific papers; and spent countless hours perusing space mission websites, which contain a dizzying amount of information, imagery, and maps on space vacation destinations near and far. For a complete list of our sources, please visit guerilla-science.org/intergalacticsources.

At nasa.gov, you will find endless information on your favorite getaway spot in the solar system and beyond. For those interested in doing a deep dive into technical and scientific reports about space exploration, https://www.sti.nasa.gov/ is a good start. If you have a favorite place you want to keep up with, here are the websites for recent missions:

THE MOON: lunar.gsfc.nasa.gov
MERCURY: messenger.jhuapl.edu
VENUS: global.jaxa.jp/projects/sat/planet_c
MARS: mars.nasa.gov

REFERENCES

JUPITER: missionjuno.swri.edu
SATURN: saturn.jpl.nasa.gov
PLUTO: pluto.jhuapl.edu

If Neptune and Uranus have won your heart and you're eager to keep up with the latest news from those planets, we regret to inform you that it's been a long time since they've been visited on holiday—or any day. An archive of photos from the last mission to visit, *Voyager 2*, is kept at http://voyager.jpl.nasa.gov. If you're wondering what's been going on with Neptune and Uranus since the late eighties, you'll have to stick to Earth-based observations or call your government representative to tell them you'd like to see a long-overdue mission.